GEOTECHNICAL SPECIAL PUBLICATION NO. 63

Design with Residual Materials

Geotechnical and Construction Considerations

Proceedings of sessions sponsored by
The Geo-Institute of the American Society of Civil Engineers
in conjunction with the ASCE National Convention

November 10-14, 1996

EDITED BY **Gordon Matheson**

Published by

345 East 47th Street
New York, New York 10017-2398

Abstract:

Residual materials are in-place soil and rock like material derived from the chemical weathering of rock. The nature of residual material is a function of the type of parent material and the extent the weathering process has disintegrated this material. The uses of residual materials are varied. These materials, due to their both soil and rock like properties, pose significant engineering and design challenges. This proceedings, *Design with Residual Materials: Geotechnical and Construction Considerations*, is a compilation of papers that discuss the origin of residual materials, and design and construction with residual materials. The contributors are both geologist and engineers that have extensive practical experience with these materials.

Library of Congress Cataloging-in-Publication Data

Design with residual materials : proceedings of sessions sponsored by the ASCE Geotechnical Engineering Division in conjunction with the ASCE National Convention, November 10-14, 1996 / edited by Gordon Matheson.
   p.   cm. -- (Geotechnical special publication ; no. 63)
   ISBN 0-7844-0207-8
   1. Residual materials (Geology) 2. Engineering geology. I. Matheson, Gordon. II. American Society of Civil Engineers. Geotechnical Engineering Division. III. ASCE National Convention (1996 : Washington, D.C.) IV. Series
TA709..5.D47 1996   96-36649
624.1'51--dc20   CIP

The Society is not responsible for any statements made or opinions expressed in its publications.

Photocopies. Authorization to photocopy material for internal or personal use under circumstances not falling within the fair use provisions of the Copyright Act is granted by ASCE to libraries and other users registered with the Copyright Clearance Center (CCC) Transactional Reporting Service, provided that the base fee of $4.00 per article plus $.25 per page is paid directly to CCC, 222 Rosewood, Drive, Danvers, MA 01923. The identification for ASCE Books is 0-7844-0207-8/96/$4.00 + $.25 per copy. Requests for special permission or bulk copying should be addressed to Permissions & Copyright Dept., ASCE.

Copyright © 1996 by the American Society of Civil Engineers,
All Rights Reserved.
Library of Congress Catalog Card No: 96-36649
ISBN 0-7844-0207-8
Manufactured in the United States of America.

GEOTECHNICAL SPECIAL PUBLICATIONS

1) TERZAGHI LECTURES
2) GEOTECHNICAL ASPECTS OF STIFF AND HARD CLAYS
3) LANDSLIDE DAMS: PROCESSES, RISK, AND MITIGATION
4) TIEBACKS FOR BULKHEADS
5) SETTLEMENT OF SHALLOW FOUNDATION ON COHESIONLESS SOILS: DESIGN AND PERFORMANCE
6) USE OF IN SITU TESTS IN GEOTECHNICAL ENGINEERING
7) TIMBER BULKHEADS
8) FOUNDATIONS FOR TRANSMISSION LINE TOWERS
9) FOUNDATIONS AND EXCAVATIONS IN DECOMPOSED ROCK OF THE PIEDMONT PROVINCE
10) ENGINEERING ASPECTS OF SOIL EROSION, DISPERSIVE CLAYS AND LOESS
11) DYNAMIC RESPONSE OF PILE FOUNDATIONS— EXPERIMENT, ANALYSIS AND OBSERVATION
12) SOIL IMPROVEMENT - A TEN YEAR UPDATE
13) GEOTECHNICAL PRACTICE FOR SOLID WASTE DISPOSAL '87
14) GEOTECHNICAL ASPECTS OF KARST TERRAINS
15) MEASURED PERFORMANCE SHALLOW FOUNDATIONS
16) SPECIAL TOPICS IN FOUNDATIONS
17) SOIL PROPERTIES EVALUATION FROM CENTRIFUGAL MODELS
18) GEOSYNTHETICS FOR SOIL IMPROVEMENT
19) MINE INDUCED SUBSIDENCE: EFFECTS ON ENGINEERED STRUCTURES
20) EARTHQUAKE ENGINEERING & SOIL DYNAMICS (II)
21) HYDRAULIC FILL STRUCTURES
22) FOUNDATION ENGINEERING
23) PREDICTED AND OBSERVED AXIAL BEHAVIOR OF PILES
24) RESILIENT MODULI OF SOILS: LABORATORY CONDITIONS
25) DESIGN AND PERFORMANCE OF EARTH RETAINING STRUCTURES
26) WASTE CONTAINMENT SYSTEMS: CONSTRUCTION, REGULATION, AND PERFORMANCE
27) GEOTECHNICAL ENGINEERING CONGRESS
28) DETECTION OF AND CONSTRUCTION AT THE SOIL/ROCK INTERFACE
29) RECENT ADVANCES IN INSTRUMENTATION, DATA ACQUISITION AND TESTING IN SOIL DYNAMICS
30) GROUTING, SOIL IMPROVEMENT AND GEOSYNTHETICS
31) STABILITY AND PERFORMANCE OF SLOPES AND EMBANKMENTS II (A 25-YEAR PERSPECTIVE)
32) EMBANKMENT DAMS-JAMES L. SHERARD CONTRIBUTIONS
33) EXCAVATION AND SUPPORT FOR THE URBAN INFRASTRUCTURE
34) PILES UNDER DYNAMIC LOADS
35) GEOTECHNICAL PRACTICE IN DAM REHABILITATION
36) FLY ASH FOR SOIL IMPROVEMENT
37) ADVANCES IN SITE CHARACTERIZATION: DATA ACQUISITION, DATA MANAGEMENT AND DATA INTERPRETATION
38) DESIGN AND PERFORMANCE OF DEEP FOUNDATIONS: PILES AND PIERS IN SOIL AND SOFT ROCK
39) UNSATURATED SOILS
40) VERTICAL AND HORIZONTAL DEFORMATIONS OF FOUNDATIONS AND EMBANKMENTS

41) PREDICTED AND MEASURED BEHAVIOR OF FIVE SPREAD FOOTINGS ON SAND
42) SERVICEABILITY OF EARTH RETAINING STRUCTURES
43) FRACTURE MECHANICS APPLIED TO GEOTECHNICAL ENGINEERING
44) GROUND FAILURES UNDER SEISMIC CONDITIONS
45) IN-SITU DEEP SOIL IMPROVEMENT
46) GEOENVIRONMENT 2000
47) GEO-ENVIRONMENTAL ISSUES FACING THE AMERICAS
48) SOIL SUCTION APPLICATIONS IN GEOTECHNICAL ENGINEERING
49) SOIL IMPROVEMENT FOR EARTHQUAKE HAZARD MITIGATION
50) FOUNDATION UPGRADING AND REPAIR FOR INFRASTRUCTURE IMPROVEMENT
51) PERFORMANCE OF DEEP FOUNDATIONS UNDER SEISMIC LOADING
52) LANDSLIDES UNDER STATIC AND DYNAMIC CONDITIONS - ANALYSIS, MONITORING, AND MITIGATION
53) LANDFILL CLOSURES — ENVIRONMENTAL PROTECTION AND LAND RECOVERY
54) EARTHQUAKE DESIGN AND PERFORMANCE OF SOLID WASTE LANDFILLS
55) EARTHQUAKE-INDUCED MOVEMENTS AND SEISMIC REMEDIATION OF EXISTING FOUNDATIONS AND ABUTMENTS
56) STATIC AND DYNAMIC PROPERTIES OF GRAVELLY SOILS
57) VERIFICATION OF GEOTECHNICAL GROUTING
58) UNCERTAINTY IN THE GEOLOGIC ENVORONMENT
59) ENGINEERED CONTAMINATED SOILS AND INTERACTION OF SOIL GEOMEMBRANES
60) ANALYSIS AND DESIGN OF RETAINING STRUCTURES AGAINST EARTHQUAKES
61) MEASURING AND MODELING TIME DEPENDENT SOIL BEHAVIOR
62) CASE HISTORIES OF GEOPHYSICS APPLIED TO CIVIL ENGINEERING AND PUBLIC POLICY
63) DESIGN WITH RESIDUAL MATERIALS; GEOTECHNICAL AND CONSTRUCTION CONSIDERATIONS

PREFACE

Residual materials are in-place soil and rock like material derived from the chemical weathering of rock. The nature of residual material is a function of the type of parent material and the extent the weathering process has disintegrated the parent rock. The uses of residual materials are varied. These materials, due to their both soil and rock like properties, pose significant engineering and design challenges.

This document is a compilation of papers that discuss the origin of residual materials and design and construction in residual materials. The contributors are both geologist and engineers that have extensive practical experience with these materials.

These papers were presented at the ASCE Annual Convention in Washington D.C. in November, 1996. Each paper was reviewed by three peer reviewers and received positive reviews from all reviewers.

In accordance with the practice of the Geotechnical Engineering Division, all papers published in conference proceedings must undergo peer review prior to being accepted. The standards for the peer review are the same as those standards for papers reviewed for possible publication in the ASCE *Journal of Geotechnical Engineering*. For acceptance, each paper was required to receive two positive reviews. Mandatory revisions must be made prior to final acceptance and publication. All papers published in this volume are eligible for discussion in the ASCE *Journal of Geotechnical Engineering* and for ASCE awards.

This symposium was a joint effort between ASCE Engineering Geology Committee and the Association of Engineering Geologists. We would like to thank Mr. Paul Burkhart, Mr. Eric Rehwoldt and Mr. Allen Cadden who provided peer review of the papers submitted. We would also like to thank the staff at ASCE for assisting in production of this publication.

The Editor,
Gordon Matheson
Bethesda, Maryland

CONTENTS

Appalachian Piedmont Regolith: Relations of Saprolite and Residual Soils to Rock-Type
    Milan J. Pavich 1

Predicting the Mode, Susceptibility, and Rate of Weathering of Shales
    Paul M. Santi and Engin C. Koncagül 12

Embankment Dams in the Piedmont/Blue Ridge Province
    Chuck Wilson and Ray Martin 27

In Situ Measurement of Rockfill Properties
    Anne Eckert Clift 37

Rethinking Foundation Design in Karst Residuum
    Raymond A. DeStephen and Steven E. Conner 49

Applications of Soil Nailing in Residual Soil
    James W. Sigourney 57

Estimation of In Situ Hydraulic Properties of Saprolite
    Gordon M. Matheson 66

Subject Index 75

Author Index 76

Appalachian Piedmont Regolith: relations of
saprolite and residual soils to rock-type

Milan J. Pavich
U.S. Geological Survey
MS 955 Reston, VA 22092

Abstract

Saprolite is a major product of rock weathering on the Appalachian Piedmont from New Jersey to Alabama. On the Piedmont, it is the primary substrate from which residual soils are developed. Properties of saprolite and residual soils are highly related to their parent rocks. Studies of cores and outcrops illustrate that rock structure and mineralogy control upland regolith zonation. Saprolite develops by in situ chemical alteration of a wide variety of mafic to highly silicic rocks. Thickness of upland saprolite varies from a few meters on mafic rocks to tens of meters on silicic rocks. Saprolite thickness decreases with increasing slope and saprolite is generally thin or absent in valley bottoms.

Massive residual subsoils and soils develop by physical and chemical processes that alter the upper few meters of saprolite. The fabric, texture and mineralogy of residual soils are distinctly different from underlying saprolite. The boundary between soil and saprolite is often gradual, and often a zone of low permeability.

Geologic maps are useful guides to Piedmont regolith thickness and zonation. In regional design studies, geologic maps and regolith characteristics can be useful in environmental decision-making.

Introduction

From a surficial geologic perspective, saprolite is the most important residual material on the Appalachian Piedmont (fig. 1). The Piedmont is a large area of low to moderate relief landscape developed on metamorphic and igneous rocks of the Appalachians. Saprolite comprises over 75% of the regolith (i.e. unconsolidated material overlying bedrock) volume on the

---

Geologist, U.S. Geologic Survey, Reston, Virginia

Piedmont, and constitutes the major zone of shallow water movement and mineral alteration. Saprolite, the residual product of isovolumetric weathering of parent rocks, underlies and can be distinquished from residual soils and subsoils which have lost the parent-rock fabric preserved in saprolite. Residual soils comprise the pedogenic A and B horizons defined in soil taxonomy (Soil Survey Staff, 1975). Recent publications, such as by Cremeens et al. (1994) have emphasized the need for better information about sub-soil regolith properties. Saprolite provides a good opportunity for systematic investigation of regolith because of the relation of saprolite properties to parent bedrock and the vertical zonation produced during isovolumetric weathering.

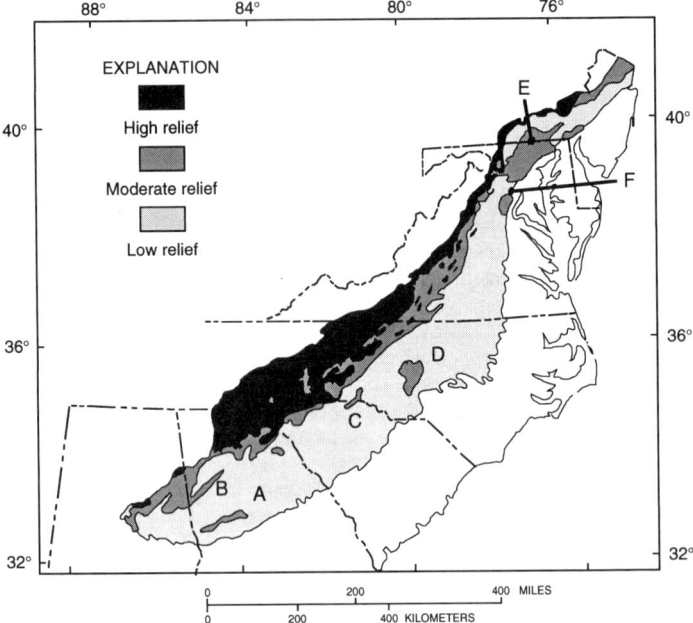

Figure 1. Generalized relief map of the Appalachian Piedmont and Blue Ridge provinces from Hack (1980). Relief classes are: low, less than 90-120m; moderate, 120-240m; high, greater than 240m and as high as 1200m. Saprolite is most common in low to moderate relief areas of the Piedmont although it also develops in high relief areas of the Blue Ridge. A=Pine Mountain, GA; B=Brevard Zone; C=Kings Mountian Belt; D=Uwharrie Mountians; E=Westminster Highland; and F=Fairfax County, VA.

Selected engineering properties of saprolite on different rock types, modified from Obermeier and Langer (1986) are shown in Table 1. This summary shows that there are significant variations in saprolite thickness and mechanical properties related to rock type. This paper will not explore the reasons for those variations. A detailed treatment of the controls on saprolite properties is found in Pavich et al. (1989). Distinct differences between residual soils and saprolites from which they are derived are expressed in engineering and hydrologic properties.

Regolith/Landscape Relations to Parent Rock-Type

Typical foliated, felsic quartz- and mica-rich metamorphic rocks produce a relatively uniform distribution of saprolite thickness with respect to landscape position. Figure 2 shows the distribution of residual soil and saprolite in a valley cross section on a weathered, metamorphosed granite in northern Virginia from Pavich (1986). Similar landscape relations have been documented in North Carolina by Daniels et al. (1984) and other parts of the Piedmont. The major features observed are that the saprolite is thickest, up to 20m, beneath upland drainage divides, that it thins along slopes to a minimum along steep valley side-slopes. The residual soil, developed *in situ* from saprolite, is usually less than a meter thick in all positions, and colluvium of varying thickness occupies steeper valley slopes.

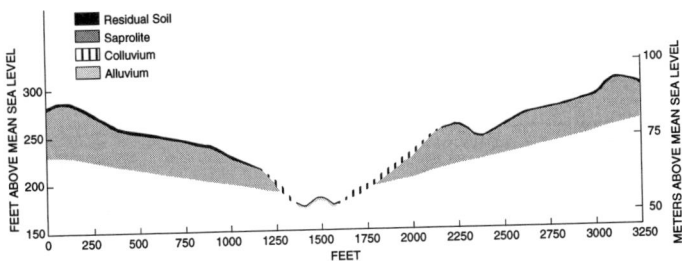

Figure 2. Valley cross section from Pavich (1986) showing the distribution of residual soil, saprolite, colluvium and alluvium in a small tributary to the Occoquan River, VA. Saprolite is thickest beneath uplands and thins along valley slopes.

On mafic rocks, such as diabase, and ultramafic rocks, such as sepentinite, the saprolite is thinner than on rocks containing resistant skeletal minerals such as quartz and muscovite. Figure 3 shows the contrast between

Table 1. Saprolite Properties

Rock Type	Fabric	Unified Soil Classification	Suitability as compacted material	Excavation Properties	Allowable Bearing Pressure, kgf/cm2
Ultramafic Serpentinite	Massive to locally foliated and tightly folded	Commonly CH in upper 70cm	Commonly unsuitable because of plastic clay	Light to moderate power equipment for plastic soil	Commonly 1-2 in shallow saprolite. Unweathered bedrock can support very large loads close to ground surface.
Mafic "Greenstone" Metagabbro Metadiorite	Massive to foliated	B-horizon: CH, MH-CH, MH Saprolite: MH, ML, SM	Commonly unsuitable on massive rocks because of plastic clay and thin saprolite. Saprolite on foliated rocks moderately easy to compact	Light power equipment suitable to 2-3m. Massive weathered rock must be blasted	Commonly 1-2 in shallow saprolite. Possible shrink/swell problems in upper 2m.
Diabase	Intrusive dikes and sills	B-horizon: CH MH-CH, CL Saprolite: MH, ML, SM-SC, SC	B-horizon highly plastic, corestones common in saprolite	Light power equipment to 3m, weathered rock must be blasted	Commonly 1-2 in shallow saprolite. Possible shrink-swell problems
Granitoid Rocks Granite, Granodiorite, pegmatite	Massive to foliated, vertical foliation common along boundaries of igneous bodies	B-horizon: MH-CH, CL, SC, SM Saprolite: SC, SM	Saprolite generally compacts well, generally strong and not highly compressible	Light power equipment to >3m, increasingly difficult with depth. Quartz veins may require blasting.	Commonly 1-2 in shallow saprolite, allowable load increases with depth.
Gneiss, schist, and metagreywacke	Massive to well-foliated rocks,	Saprolite: ML, SM (MH)	Saprolite generally compacts well, ease of compaction depends upon moisture content	Light power equipment to 3m, increasing difficulty with depth	Commonly 1-2 in shallow saprolite, values increase rapidly with depth
Quartz bodies	Veins, dikes and pods in schist, gniess and granite	Not applicable	Not applicable	Power equipment for smaller bodies, wider veins may require blasting	Not applicable

regolith profiles on felsic and mafic rocks. Landscapes underlain by mafic rocks tend to have thin regolith on upland positions. On ultramafic rocks, saprolite is generally absent and the soil is thin or absent.

There are important exceptions to this general relation to bedrock type. In North Carolina, for example, a large area of felsic volcanics known as the Carolina Slate Belt typically displays very low relief and thin saprolite. By comparison with other metasedimentary rocks of the Piedmont, the slates tend to have near horizontal bedding and foliation which may limit the depth of weathering. Two other important exceptions are:

1) quartz veins, particularly in intrusive granitod rocks. Quartz veins can be a few cm to a few meters in width. The thinner ones are common in saprolite and often produce a surficial lag of quartz gravel on upland surfaces. Larger veins may not exhibit relief at the soil surface, but can cause excavation problems.

2) massive and sheeted rocks. Granitic intrusives, such as Stone Mountain, Georgia, are good examples of lithologies that are resistant to weathering and erosion because of massive structure. Granite "balds" with less relief than Stone Mountain are common on the Piedmont and cover areas that are easily identified on geologic maps. Massive greenstone, a metavolcanic rock, and diabase, a mafic intrusive rock, can also be relatively resistant to weathering and produce local topographic highs.

The characteristics noted for the various rock types can be used as a rough guide to what can be expected in the field setting. In areas where detailed geologic maps (e.g. 1:24,000 scale or larger) are available, the major rock types noted in the key should provide an initial guide to the subsurface condition that might be encountered. Because of the scale of variation in rock type in tightly folded metamorphic rocks, engineering properties can vary significantly at the scale of an engineered structure.

Saprolite versus Soil

The soil/saprolite boundary is one of the most important transitions with respect to mechanical and hydrologic properties. Figure 3 shows a comparison of soil and saprolite thicknesses on foliated metasedimentary rocks and mafic igneous rocks. In the former, the saprolite is thick, there is a transitional "massive" soil zone and a residual soil (pedogenic B and A horizons) that is generally less than 1m thick. In the mafic rock, the saprolite is generally thin (<2m) and there is a less gradual transition from saprolite to soil.

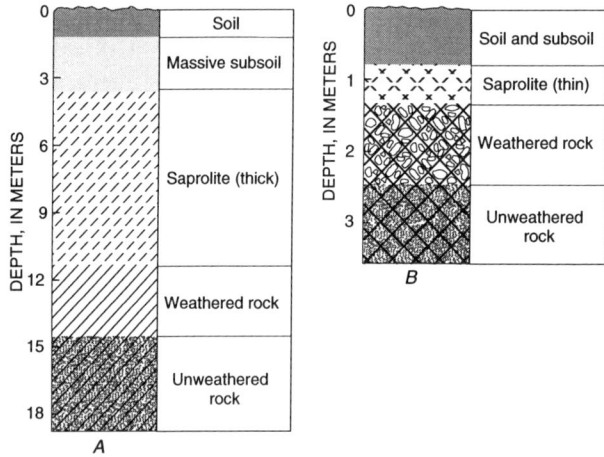

Figure 3. Generalized weathering profiles developed on crystalline rocks of the Piedmont in Fairfax County, VA (from Pavich et al., 1989). A, Foliated metasedimentary and granitic rocks. B, Massive igneous rocks such as diabase.

The transition from saprolite to soil is based on the dominant processes by which they are generated. Figure 4 summarizes the processes that determine the thickness and transitions between the regolith zones. Table 2 describes the major zones in vertical sequence and some general properties. From the base of the profile upward, the weathered rock zone is structurally identical with the parent rock but has undergone chemical alteration due to oxidation and hydration reactions. The boundary between weathered and unweathered rock, the weathering front, is generally very sharp and determines refusal with respect to split spoon sampling.

The saprolite is distinquished from the weathered rock by its clay content and much weaker consistency, or strength. Although the saprolite retains the structure and fabric of the parent rock, due to isovolumetric weathering reactions, it is generally very friable and easy to excavate. In situations where the saprolite exceeds a few meters in thickness, generally beneath uplands on foliated quartzofeldspathic rocks, the upper part of the saprolite is a skeletal remnant of the more chemically resistant parent minerals, usually quartz and muscovite. The transition from saprolite to massive subsoil is marked by loss of the original parent rock fabric, rearrangement of resistant minerals, and reduction of volume. The soil is distinct from the massive zone by virtue of B-horizon structure, clay content, color and evidence of more intense mineral alteration than subjacent zones.

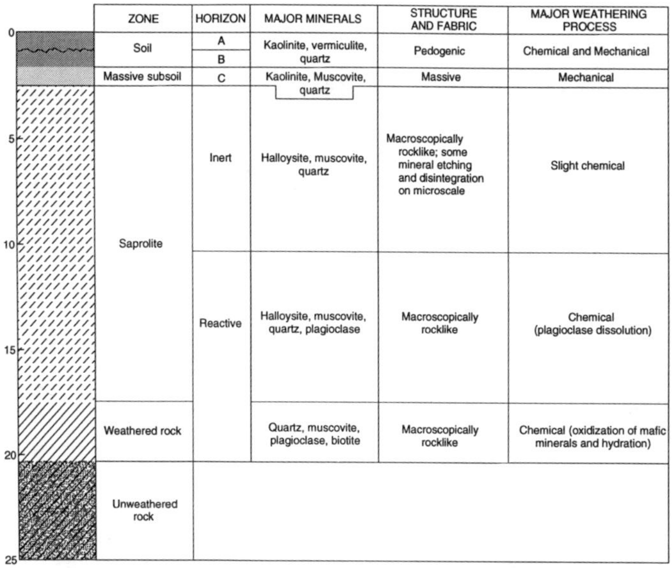

ZONE	HORIZON	MAJOR MINERALS	STRUCTURE AND FABRIC	MAJOR WEATHERING PROCESS
Soil	A	Kaolinite, vermiculite, quartz	Pedogenic	Chemical and Mechanical
	B			
Massive subsoil	C	Kaolinite, Muscovite, quartz	Massive	Mechanical
Saprolite	Inert	Halloysite, muscovite, quartz	Macroscopically rocklike; some mineral etching and disintegration on microscale	Slight chemical
	Reactive	Halloysite, muscovite, quartz, plagioclase	Macroscopically rocklike	Chemical (plagioclase dissolution)
Weathered rock		Quartz, muscovite, plagioclase, biotite	Macroscopically rocklike	Chemical (oxidization of mafic minerals and hydration)
Unweathered rock				

Figure 4. Generalized weathering profile for thick regolith developed on upland, quartzofeldspathic metasedimentary rocks (from Pavich et al., 1989).

Soil/Saprolite Hydrologic Relations

The transition between soil and saprolite is termed the massive zone in Pavich et al. (1989). This zone exhibits the alteration of saprolite into a material that has some, but not all, soil properties. Based on thin-sections, there is a physical reorganization of skeletal minerals originally in foliation planes such as muscovite and quartz into an unoriented matrix of skeletal minerals and finer matrix (Pavich et al., 1989). The permeability of the massive zone is often significantly less than that of the soil or saprolite.

Amoozegar and Hoover (1989) documented the movement of water through drainfields constructed on soil and saprolite. Field measurements of saturated hydraulic conductivity (K_S) showed that K_S decreased to a

Table 2. Description of weathering profile for igneous and metasedimentary rocks in Fairfax County, VA. (modified from Deere and Patton, 1971)

Zone	Description[1]	RQD[2] (NX core percent)	Percent core recovery (NX core)	Relative permeability	Relative strength	Common thickness (meters)
Soil — A Horizon	Top soil, roots, organic material. Zone of leaching and eluviation. Generally porous and sandy.	Not applicable.	0	Medium to high	Very low	0.1–0.2
B Horizon	Characteristically clay enriched, also accumulations of Fe and Al. No relict structures present.	Not applicable.	0	Low	Commonly low, medium if very dry.	0.3–1.0
Massive subsoil	No relict rock structure. Less dense than soil B horizon. Less clay than soil B horizon. Depleted in cations and silica relative to Fe and Al. May contain clasts of saprolite.	Not applicable.	0	Medium	Low	0.5–1.0
Saprolite	Relict rock structures retained. Clay-bearing silt or clay-bearing sand grading to sand at depth. Commonly micaceous; feldspars and mafic minerals altered to clays. Less than 10 percent core stones. Joints strongly cemented with oxides at many places.	0 or not applicable.	Generally 0–10 percent.	Medium	Low to medium (relict structures very significant.	1–15
Weathered rock — Transition (from saprolite to partly weathered rock).	Highly variable, saprolitelike to rocklike. Fines commonly line to coarse sand (grus). 10–95 percent core stones. Feldspars and mafic minerals altered in part.	Variable, generally 0–50 percent.	Variable, generally 10–90 percent.	High (water losses common during drilling).	Medium to low where weak structures and relict structures are present.	0.3–3
Partly weathered rock.	Rocklike, soft to hard rock. Joints stained to altered. Slight alteration of feldspars and mafic minerals.	Generally 50–75 percent.	Generally 90 percent.	Medium to high	Medium to high[3]	0.3–3
Unweathered rock	No iron stains to trace along joints. No weathering of feldspars and micas. No sheared zones.	>75 percent (generally >90 percent.	Generally 100 percent.	Low to medium	Very high[3].	—

[1] The descriptions provide the only reliable means of distinguishing the zones.
[2] RQD stands for Rock Quality Designation, described in Deere and others (1967). RQD in percent equals length of core pieces 4 in. (10.3 cm) and longer divided by length of run times 100. NX core diameter in 1.75 in. 14.5 cm).
[3] Considering only intact rock with no adversely oriented geologic structures.

minimum of 2cm/day at 75 to 100 cm below the soil surface and increased below that depth. The minimum appears to correspond to the upper part of the massive zone, or Bt3 horizon, below the horizon of maximum clay content, the soil Bt2 horizon. They noted (p. 112) that:

" The Bt3 represented a horizon that was clayey and had less well developed soil structure than above, and did not have the well preserved root channels and remnant foliation planes that were present in the Bt/C horizon below it."

The presence of a low permeability zone at the soil/saprolite transition may be important to design of drainfields and other shallow structures. In situations where perching of soil water may be of concern, it may be of importance to excavate and inspect the depth of the lower soil/saprolite transition.

Saprolite Taxonomy

One of the impediments to systematic investigation of, and communication about, regolith and saprolite is the lack of a simple taxonomic system. Recently, Buol and others (1994) have proposed a taxonomic system for saprolite and related transported regolith such as alluvium, colluvium, and petrosediments. Since saprolite properties are different from soil properties, the soil taxonomic system (Soil Survey Staff, 1975) is not sufficient for describing or classifiying saprolite. One of the underlying concepts of the taxonomic system is that saprolite characteristics are similar over large areas. Thus data bases developed for specific sites should have value when applied to other sites.

Buol et al. (1994) used material strength in their definition of saprolite-regolith:

> "Saprolite -regolith materials have unconfined compressive strength less than 100MPa, and are either not penetrated by plant roots, except at intervals greater than 10 cm, or occur more than 200 cm below the soil surface, whichever is shallower. The lower depth limit of saprolite-regolith material to be classified in this system is not specified, except that it will not include material with unconfined compressive strength of 100Mpa or greater."

The proposed classification system is an approximation that will be amended and modified by use. It provides an imperfect, but common language for scientists and engineers to communicate about regolith. Research on saprolite and related regolith still falls in the grey-area between pedology and geology. The expanded use of saprolite-regolith in engineering and waste management makes it important to improve our common and scientific knowledge about these materials.

Geologic Maps, Saprolite and Land Use Decisions

The relations of regolith characteristics to bedrock and topographic position are well expressed on the Piedmont. Geologic maps, therefore, provide useful guides to the expected subsurface conditions in areas of saprolitic weathering. Recent applications of GIS, such as siting of a Piedmont landfill (Bernknopf et al., 1993), illustrate that geologic maps and subsurface information can be beneficially applied to environmental problems. When combined with economic data long-term costs of environmental decisions can be reduced. Geologic maps and map-based data bases can provide valuable information for hillslope and floodplain development, design and construction of surface water impoundments, siting of municipal well fields, etc. (Bernknopf et al., 1993). The zonal properties of saprolite provide a useful framework for the construction of three-dimensional data bases in deeply weathered terrain. Thus, geologic maps and data bases of regolith characteristics can be useful initial guides to decision making.

Conclusion

Regolith of the low-to-moderate relief Appalachian Piedmont is composed dominantly of saprolite and residual soil developed from saprolite. The thickness of regolith is related both to position on slope and to parent rock type. Geologic and topographic maps can provide a rough guide to expected regolith thickness. The soil/saprolite boundary is an important transition in mechanical and hydrologic properties. This transitional zone is often characterized by low permeability which can be of importance to subsurface drainage. Systematic relations of saprolite and residual soils to bedrock and topography provides a basis to developing regional data bases and a taxonomic code for saprolite and related materials. Such data bases can be used effectively in GIS analysis of land-use.

References

Amoozegar, Aziz, and Hoover, M.T., 1989, Movement of water and chemical pollutants from wastewater disposal systems through the soil and saprolite of Piedmont landscapes, Water Resources Research Institute, University of North Carolina, Report No. 249, 153p.

Bernknopf, R.L., Brookshire, D.S., Soller, D.R., McKee, M.J., Sutter, J.F., Matti, J.C., and Campbell, R.H., 1993, Societal Value of Geologic Maps, U.S. Geological Survey Circular 1111, 53p.

Buol, S.W., 1994, Saproilite-regolith taxonomy-an approximation, in Cremeens D.L. et al., eds., Whole Regolith Pedology, Soil Science Society of America Special Publication No. 34, p. 119-132.

Cremeens, D.L., Brown, R.B., and Huddleston, J.H., eds., 1994, Whole Regolith Pedology, Soil Science Sciety of America Special Publication 34, Madison, WI, 136p.

Daniels, R.B., Kleiss, H.J., Buol, S.W., Byrd, H.J., and Phillips, J.A., 1984, Soil Systems in North Carolina, Bulletin 467, North Carolina Agricultural Research Service, 77p.

Deere, R.U. and Patton, F.D., 1971, Slope stability in residual soils: 4th Panamerican Conference on Soil Mechanics and Foundation Engineering, p. 87.

Hack, J.T., 1980, Rock control and tectonism - their importance in shaping the Appalachian Highlands, U.S. Geological Survey Professional Paper, 1126-B, p. 17.

Lambe, T.W., 1951, Soil testing for engineers: New York, John Wiley, 165p.

Lambe, T.W. and Whitman, R.V., 1969, Soil Mechanics: New York, John Wiley, 553p.

Obermeier, S.F., and Langer, W.H., 1986, Relationships between geology and engineering characteristics of soils and weathered rocks of Faifax County and vicinity, Virginia, U.S. Geological Survey Professional Paper 1344, 30p.

Pavich, M.J., 1986, Processes and rates of saprolite production and erosion on a foliated granitic rock of the Virginia Piedmont, in Colman, S.M. and Dethier, D.P., Rates of Chemical Weathering of Rocks and Minerals, Academic Press, Orlando, p. 552-590.

Pavich, M.J., Leo, G.W., Obermeier, S.F., and Estabrook, J.R., 1989, Investigations of the characteristics, origin, and residence time of the upland residual mantle of the Piedmont of Fairfax County, Virginia, U.S. Geological Survey Professional Paper 1352, 58p.

Soil Survey Staff, 1975, Soil Taxonomy: Washington, D.C., U.S. Department of Agriculture Handbook 436, 754p.

Terzhagi, Karl and Peck, R.P., 1948, Soil mechanics in engineering practice: New York, John Wiley, 729p.

Predicting the Mode, Susceptibility, and Rate of Weathering of Shales

Paul M. Santi[1] and Engin C. Koncagül[2]

Abstract

Fifty three shale samples from various shale units in Colorado, Missouri, and Illinois were collected and tested for slake index, jar slake, and slake durability. Sets of samples were collected from the same unit, but with different degrees of weathering. Analysis of laboratory tests and trends along these weathering profiles showed two distinct weathering behaviors, based on the mode and degree of slaking. Minor slaking of the surfaces of blocks, termed "chip" or "surface" slaking, represents a slaking process which slows down over time. This occurs because as surface layers slough off, the exposed surface area decreases as the intact sample reduces in size. Breakdown of the blocks through microfractures is termed "body" or "block" slaking and represents a slaking process that speeds up over time, as induced fractures rapidly degrade material integrity. Furthermore, the weathering process can induce changes in both modes and rates of slaking.

Introduction

Weak or shaly rocks typically weather by slaking, which is the process by which material sloughs or disintegrates upon contact with water. Slaking is an important process in engineering because it can cause rapid changes in strength and durability, leading to problems with erosion, slope stability, settlement, bearing capacity and drainage.

The purpose of this study is to show the relationship between the mode of slaking and the slaking rate and susceptibility to slaking. This relationship may be deduced using the slake index test. This test consists soaking samples in water,

---

[1] Assistant Professor, Department of Geological and Petroleum Engineering, University of Missouri - Rolla, Rolla, MO 65401
[2] Ph.D. Candidate, Department of Geological and Petroleum Engineering, University of Missouri - Rolla, Rolla, MO 65401

washing away loosened or slaked particles through a #10 sieve, and drying the retained material. The slake index is defined as the percentage of material lost.

One advantage of the slake index test is that several dry-soak-wash cycles may be performed on the same sample, to increase the rigorousness of the test.. An example progression of slake index values over five cycles is shown on Figure 1b. The test also models natural weathering processes. According to Hudec (1982) and Santi (1995b), the slake index cycles mimic the stresses a material feels in nature due to wetting and drying or freezing and thawing. A third advantage of the test is that the numerical result allows quantitative comparison of sample behavior and removes subjectivity from the analysis.

The slake index test may also be used to gauge the resistance of shales to weathering and to predict the mode of future weathering. This application of the slake index test views shales or other weak rock materials as a two-phased system, containing both intact rock-like blocks and weathered soil-like matrix. Using this model, the one-cycle slake index represents the approximate percentage of matrix material, that is the loosely-bound and water-sensitive material surrounding sample blocks. The five-cycle slake index provides a rough indication of the resistance of the blocks to weathering. The repeated drying and soaking accelerates the natural stresses the blocks would normally experience during weathering.

The progression of slake index values from one to five cycles is shown by a line connecting the individual cycle results on Figure 1b. This progression may be approximated as a single point on Figure 1a, which is a plot of one-cycle versus five-cycle values. The location of this point reveals the relative magnitude of one- and five-cycle values: points towards the upper left show high slaking rates with high five-to-one-cycle ratios, and points towards the upper right show lower slaking rates with low five-to-one-cycle ratios.

A plot such as Figure 1a can also be used to compare samples of the same material which have been weathered to different degrees. In this instance, the expected path of slaking for a series of seven samples which have been weathered to different degrees is shown. These samples may have come from various depths in the same borehole, where shallow samples have experienced significant weathering (such as sample 7), and progressively deeper samples are progressively less weathered (such as sample 1). For these materials, the slaking process is assumed to mimic the weathering process.

The samples in Figure 1 show a constant rate of slaking. That is, the same amount of material is lost in each slake cycle; therefore, the five-cycle value is five times the one-cycle value. The plot of individual slake index values is a straight line. Note that the line makes an angle along the top of the graph, where samples 5, 6, and 7 are weathered to the degree that all the material passes the sieve before five slake index cycles have been completed.

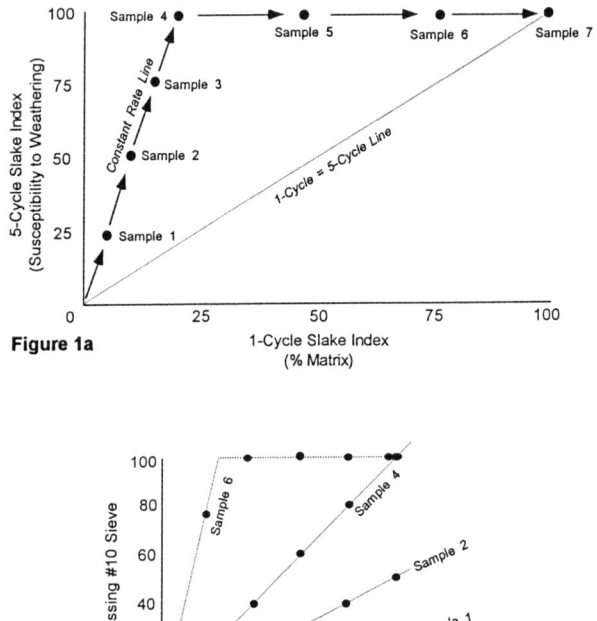

Figure 1 - Progression of Slake Index Values for a Series of Samples. Sample 1 is a fresh rock sample, and each sample in the sequence from 1 to 7 is progressively more weathered. In this case, the series of samples slakes at a constant rate, as indicated by the straight line plots on 1a. On 1b, samples which slake at a constant rate plot on the "constant rate line," along which the five-cycle value is five times the one-cycle value. Sample 4, which falls at the end of the constant rate line, has weathered enough that all the material slakes after five wet-dry cycles. Samples 5, 6, and 7 have weathered more than Sample 4, and follow a weathering path along the top of the graph. This path is the logical continuation of the constant rate line for highly weathered samples.

Figure 2 shows a series of five samples with an accelerated rate of slaking. In this case, the five-cycle value is more than five times the one-cycle value. The plot of individual slake index values is a curved line with increasing slope (Figure 2b).

Figure 3 shows a series of five samples with a decelerating rate of slaking, where the five-cycle value is less than five times the one-cycle value. The plot of individual slake index values is a curved line with decreasing slope (Figure 3b).

The advantages of this sort of analysis are that the characteristics and progression of slaking and degree of weathering are displayed on a single graph, and the engineer may logically evaluate the expected future behavior of materials through the weathered zone. This study further describes the expanded interpretation of the slake index test, and supports this interpretation using laboratory tests and microscope examination.

Background

Some researchers have proposed that slaking be used as the diagnostic behavior to distinguish between rock-like materials and soil-like materials (Morgenstern and Eigenbrod, 1974; Santi, 1995a). Others have noted that slaking occurs more strongly in compacted shales than in cemented shales (Burwell and Moneymaker, 1950) and in materials which have been completely dried rather than those retaining moisture (Mitchell, 1993). Mitchell (1993) identifies three mechanisms which lead to slaking: dispersion of soil particles, swelling caused by stress relief and water absorption, and tensile stresses resulting from compression of entrapped air as water is absorbed.

Moriwaki and Mitchell (1977) suggest four common modes of slaking:

- Dispersion slaking, where submerged samples completely dissolve, releasing air bubbles in the process, and forming a cloudy residue. They expect this mode to dominate when Na-Kaolinite is abundant.
- Swelling slaking, where samples disintegrate in response to excessive increases in volume of clay minerals such as Na-Montmorillonite.
- Body slaking, which is the macroscopic disintegration of material. In this mode, the material fractures throughout into irregular blocks. They expect this mode to dominate when Ca-Kaolinite and Ca-Illite are abundant.
- Surface slaking, where samples slake from the surface inward by development of parallel, thin plates. They suggest that this mode is linked to Ca-Montmorillonite.

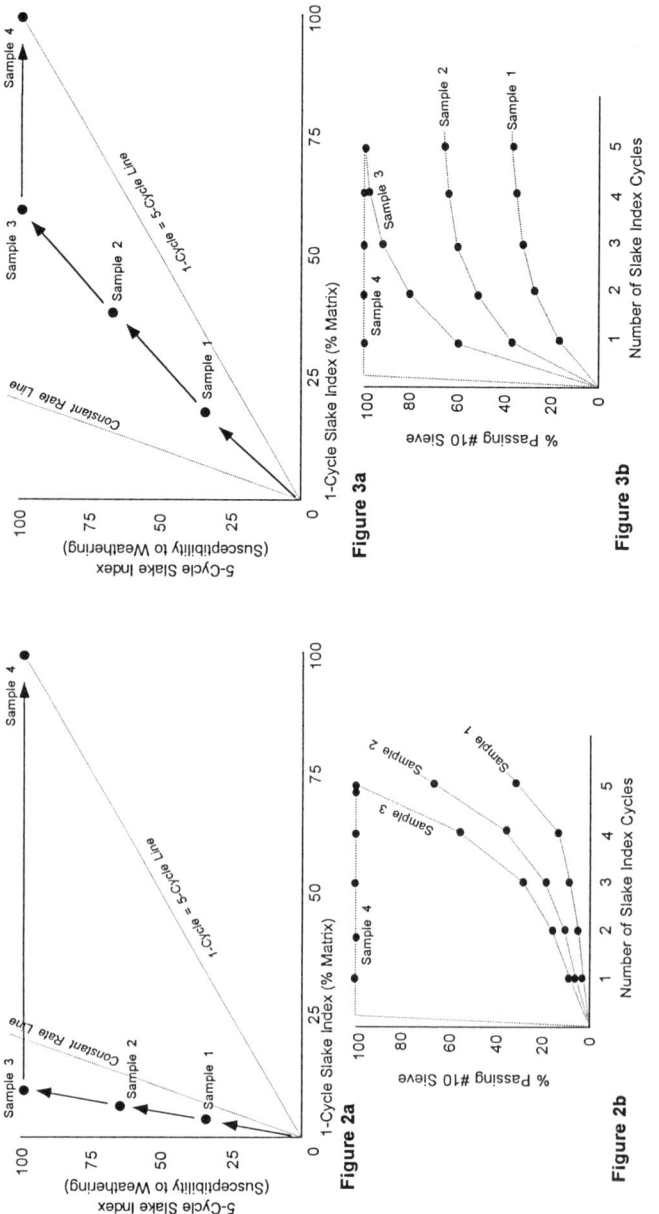

Figure 2 - Progression of Slake Index Values for Samples with an Accelerated Slaking Rate. The accelerated slaking rate is shown by the upward curving lines on the 2b. These samples plot to the left of the constant rate line on the 2a.

Figure 3 - Progression of Slake Index Values for Samples with a Decelerated Slaking Rate. The decelerated slaking rate is shown by the flattening lines on the 3b. These samples plot to the right of the constant rate line on 3a.

Perry and Andrews (1982) observed three modes of slaking, which correspond well to those observed by Moriwaki and Mitchell. Perry and Andrews also report the behavior of mine spoils subject to each slaking mode:

- Slaking to inherent grain size, which corresponds to dispersion slaking. These materials showed significant embankment stability problems and severe erosion problems.
- Chip slaking, which corresponds to surface slaking. These materials produced relatively stable embankments and had minor erosion problems.
- Block slaking, which corresponds to body slaking. These materials produced stable embankments and had minor erosion problems.

Although slaking occurs in rocks other than shales, and does not occur in every shale, it is one of the most significant processes controlling shale behavior (Burwell and Moneymaker, 1950). Slaking is largely responsible for the differences in long-term behavior between shale and other rock types. Therefore the slake index test, as a measure of the degree of slaking, and the mode of slaking, as an indicator of the processes acting internally in the sample, may be used together to describe the short-term slaking and long-term behavior of shales.

Methodology

In addition to the slake index test, several additional tests were performed to confirm the application and reliability of the slake index interpretation. The slake durability test, accepted as an industry standard to gauge reaction to water and abrasion was used to confirm the validity of slake index results. The jar slake test assigned numerical values to slaking modes and also was used to check slake index results. Finally, microscope analysis was used to confirm slaking modes.

Slake Index Test

The slake index was first proposed by Deo (1972). For this study, methodology followed Walkinshaw and Santi (1996), with minor modifications. Six sample pieces are oven-dried for 16 hours, immersed in distilled water for 24 hours, and then washed over a #10 sieve. Washed samples are dried and weighed, and the slake index is calculated as the percent weight loss through the sieve.

Slake Durability Test

The slake durability test was conducted in accordance with ASTM D4664-87, except that tap water was used instead of distilled water to accommodate the large number of samples. In summary, the ASTM procedure recommends selecting 10 sample pieces, oven-drying them, weighing, and rotating them in the slake durability drum partially immersed in water. After 10 minutes, the sample is removed, dried, and the retained sample weight is calculated. The cycle is normally repeated a second time, and the second cycle slake durability index, ID(2), is calculated as the percent weight retained.

Jar Slake Test

The jar slake test was conducted according to the procedures outlined in Wood and Deo (1975) and Walkinshaw and Santi (1996). In summary, a 30 to 50g sample is immersed in distilled water and described after 24 hours. The categories for description are:

Category	Behavior
1	Degrades to a pile of flakes or mud
2	Breaks rapidly and/or forms many chips
3	Breaks slowly and/or forms few chips
4	Breaks rapidly and/or develops several fractures
5	Breaks slowly and/or develops few fractures
6	No change

Category 1 represents dispersion slaking or slaking to inherent grain size. Categories 2 and 3 represent chip or surface slaking. Categories 4 and 5 represent body or block slaking.

Microscope Analysis

Based on the results of slake index, slake durability, and jar slake tests, several representative samples were observed under a stereoscopic microscope using a magnification of 10X and 20X with incident light, under the following conditions:

- air dried intact pieces,
- air dried pieces soaked with water from a dropper to observe any rapid slaking,
- pieces which remained following one slake index cycle, and
- pieces which remained following five slake index cycles.

While under the microscope, samples were prodded with a long pin to pry apart incipient fractures or flakes, so that the slaking mode might be better characterized.

Results

The results of laboratory testing for slake index are summarized on Table 1. This table includes five cycles of slake index values for 18 of the 53 samples collected from various shale units in Colorado, Missouri, and Illinois. These samples represent pairs or sets of samples collected from the same unit, but showing

Table 1 - Results of Laboratory Testing.

Sample	Slake Index Analysis					Ratio	Jar	Slake Durability	
	1-Cycle	2-Cycle	3-Cycle	4-Cycle	5-Cycle	5:1 cycles	Slake	ID(1)	ID(2)
Accelerated Slaking									
Kmv21	1.39	5.49	11.34	20.73	31.14	22.39	5	85.20	74.93
Kmv22	1.40	4.23	9.76	18.14	26.24	18.80	5	85.99	72.78
Km21	10.26	37.75	72.61	91.18	98.20	9.58	5	96.70	92.82
Km22	59.56	95.74	100.00	100.00	100.00	1.68	2	85.31	64.51
Jm31	0.69	1.67	3.86	6.34	9.75	14.22	6	98.21	97.05
Jm32	6.35	37.82	60.78	82.05	89.60	14.11	4	87.22	66.46
Jm33	98.57	100.00	100.00	100.00	100.00	1.01	1	0.30	0.00
Decelerated Slaking									
Kp11	78.92	79.67	80.13	81.02	81.87	1.04	1	21.67	19.55
Kp12	0.78	0.80	0.94	1.06	1.06	1.36	6	97.87	96.84
TRc51	25.62	44.91	55.44	61.63	66.59	2.60	3	88.48	73.95
TRc52	70.54	88.34	92.80	94.08	94.71	1.34	2	47.86	22.36
PPe61	0.98	2.16	3.10	3.72	4.77	4.89	6	95.02	91.57
PPe62	27.04	46.72	51.47	53.76	55.58	2.06	3	83.50	75.91
Constant or Varying Rate of Slaking									
Kc11	5.00	14.74	25.93	33.97	41.79	8.35	5	95.17	87.55
Kc12	13.47	27.67	41.59	52.95	59.71	4.43	5	89.76	68.15
Mm21	41.87	64.02	72.37	77.92	82.62	1.97	6	****	****
Mm22	11.47	20.19	28.75	37.13	44.44	3.87	4	****	****
Mm23	5.20	10.33	16.98	22.26	27.70	5.33	4	****	****

Kmv = Cretaceous Mesaverde Group (Colorado)
Km = Cretaceous Mancos Formation (Colorado)
Jm = Jurassic Morrison Formation (Colorado)
Kp = Cretaceous Pierre Formation (Colorado)
TRc = Triassic Chinle Formation (Colorado)
PPe = Pennsylvanian Eagle Formation (Colorado)
Kc = Cretaceous Colorado Group (Colorado)
Mm = Upper Mississippian Undifferentiated (Illinois)

different degrees of weathering. Also included in Table 1 are the slake durability and jar slake test results, as well as the calculated ratio of five-cycle to one-cycle values. The remaining 35 samples are not further analyzed or presented in Table 1 for one of two reasons:

1. they cluster near the corners of the graph used in this analysis (such as Figure 2a) and are difficult to interpret, or
2. the progression of weathering between samples in the set is unclear because the range of slake index values produced by the set is so narrow.

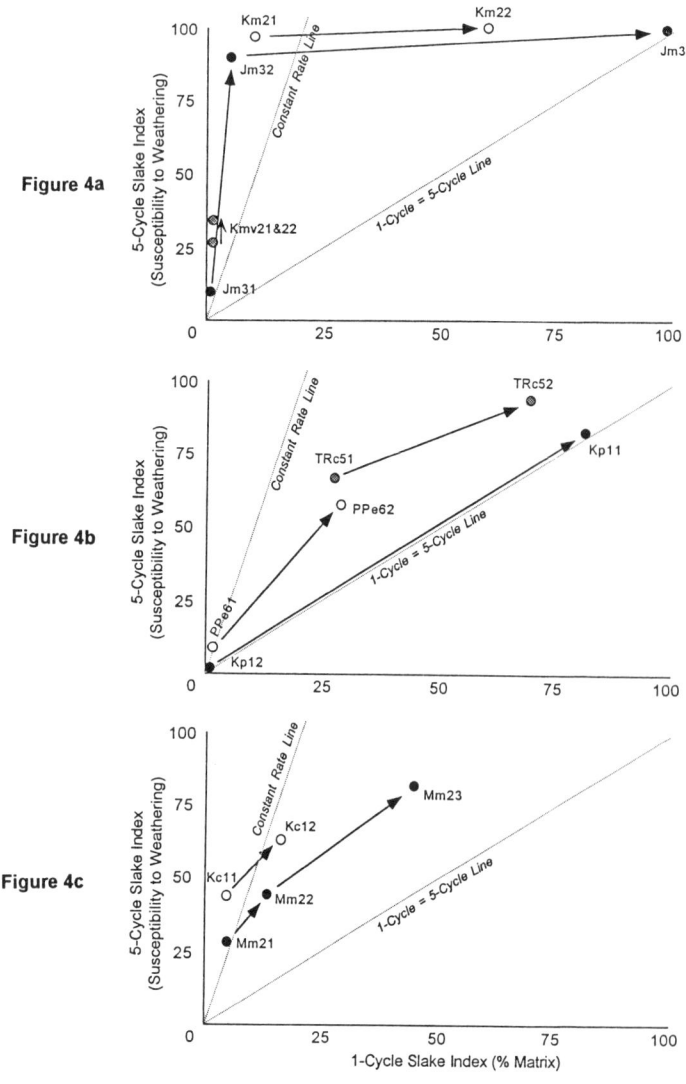

Figure 4 - Plot of Sample Sets Showing Different Rates of Slaking. Figure 4a includes sample sets with a predominantly accelerated rate of slaking, Figure 4b includes those with a decelerated rate of slaking, and Figure 4c includes those which distinctly cross the constant slaking rate line.

The data presented in Table 1 are broken into groups based on their five-cycle to one-cycle ratios. The first group of samples are those sample sets which have a clearly accelerated rate of slaking (five-cycle to one-cycle ratio is greater than five). These results are plotted on Figure 4a. Note that two samples, Km22 and Jm33 plot far to the right of the constant rate line, and have correspondingly low five-cycle to one-cycle ratios. These samples are much more highly weathered than their counterparts within the same sample set.

The second group of samples are those sample sets which have a clearly decelerated rate of slaking (five-cycle to one-cycle ratio is less than five). These results are plotted in Figure 4b.

The third group of samples in Table 1 are those sample sets which plot near the constant rate line (five-cycle to one-cycle ratio is equal to five), or which show a progression of weathering between samples in the set that crosses the constant rate line. These results are plotted in Figure 4c.

A summary of the microscope analysis of the three groups of samples included in Figure 4 is presented in Table 2. This table notes the observed slaking mode for each sample, as well as behavioral characteristics when dry, wet, or after slaking has occurred. In general, microscope observations showed that:

- samples which show accelerated weathering in the slake index test predominantly show body or block slaking,
- samples which show decelerated weathering in the slake index test predominantly show chip or surface slaking,
- the samples selected as representative of varying rates of weathering show body or block slaking,
- in spite of similar slaking modes, different samples showed different susceptibilities to "induced" fracturing or flaking by prodding and prying, and
- samples subject to dispersion slaking had rapid and complete reactions upon saturation.

In support of the validity of the slake index test as an additional method of slake analysis, linear regression correlations between slake index tests and slake durability and jar slake are included in Figure 5. These correlations were measured using all 53 samples, and not just those presented in Tables 1 and 2. Based on the correlation coefficients, the slake durability test is more closely related to the one-cycle slake index test ($R^2 = 0.7318$) than to the five-cycle slake index test ($R^2 = 0.5714$). Similarly, the jar slake test is more closely related to the one-cycle slake index test ($R^2 = 0.7631$) than to the five-cycle slake index test ($R^2 = 0.5362$).

Table 2 - Summary of Slaking Mode and Microscope Observations.

Sample	Slaking Mode	Microscope Observations
Accelerated Slaking		
Kmv21	Body or Block	When dry, sample is hard and difficult to flake. When saturated, sample is easier to fracture than when dry, but still difficult to flake.
Kmv22	Body or Block	Sample is slightly easier to fracture and flake when saturated than the Kmv21 sample. Obvious air bubbles are created in fractures and under flakes when saturated.
Km21	Body or Block	Some water absorbed, some pools on surface.
Km22	Chip or Surface	Dry sample disintegrates under finger pressure and is highly fractured.
Jm31	None	Soaking creates some small flakes parallel to surface, difficult to pry up new flakes.
Jm32	Body or Block	Soaking creates some small flakes parallel to surface, some fractures near sample edges. Easier to pry up new flakes than sample Jm31.
Jm33	Dispersion	Honeycombed, porous appearance when dry. Immediate and complete dispersion upon saturation.
Decelerated Slaking		
Kp11	Dispersion	Air bubbles released during saturation.
Kp12	None	No reaction to water, remnant fractures healed with calcite infilling.
TRc51	Chip or Surface	Sample disintegrates upon soaking.
TRc52	Chip or Surface	Air bubbles released during saturation, sample disintegrates.
PPe61	None	Sample parts readily along bedding planes, but water does not seem to enhance these partings.
PPe62	Chip or Surface	No visible fractures when sample is wet or dry. The sample is difficult to flake when dry, but the edges flake readily when wet.
Constant or Varying Rate of Slaking		
Kc11	Body or Block	When dry, fractures are apparent but cannot be easily pried apart. Fractures still cannot be pried apart when saturated.
Kc12	Body or Block	When dry, fractures can be pried apart with some difficulty. When saturated, fractures are readily pried apart.
Mm21	None	Although a few flakes appeared following saturation, it was difficult to pry up new flakes or fractures. No fractures appeared after saturation.
Mm22	Body or Block	Many fractures appeared after saturation. They were easy to separate, but the separated chunks were difficult to break or flake.
Mm23	Body or Block	Similar to samples Mm22, separated chunks were easier to flake.

Discussion

The results presented above may indicate two items which need further discussion: first, that the mode of slaking may be readily deduced using field and laboratory observations, and second, that the mode of slaking may be further interpreted to show how a sample weathers.

The mode of slaking is most easily observed using the jar slake test. Weak slaking is indicated by body or block slaking, moderate slaking is indicated by chip or surface slaking, and strong slaking is indicated by dispersion slaking. Microscope

PREDICTING WEATHERING OF SHALES 23

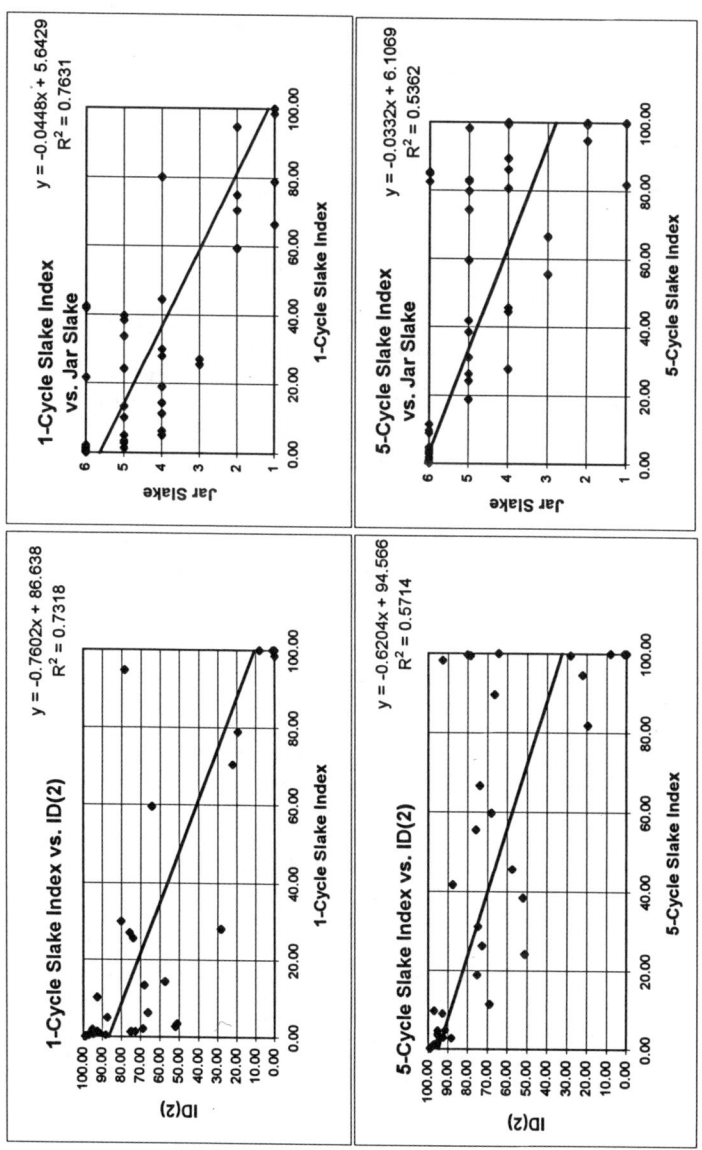

Figure 5 - Linear Regression Correlations Between Laboratory Analyses.

analysis indicates that degrees of behavior are apparent within each slaking mode. For example, samples Kmv21 and Kmv22 both show body or block slaking, but one sample is much easier to flake by prying than the second sample. Furthermore, air bubbles were created during saturation of the first sample and not in the second. These bubbles actively accelerate slaking.

The slake index analysis addresses the differences within each slaking mode by using the five-cycle slake index as a measure of the susceptibility to weathering. Sample Kmv21 is slightly less susceptible to weathering by slaking (as indicated by a lower one-cycle value) than is sample Kmv22, shown by the relative position of their plots on Figure 4a.

Another application of these results is the relationship between slaking mode and the rate of weathering. For example, samples Kmv21 and Kmv22 plot in the region of accelerated slaking. This region is defined to include materials whose five-cycle slake values are more than five times their one-cycle slake values. As shown on Figure 2, these samples lose progressively more material with each wet-dry cycle. Since these materials also show predominantly body or block slaking, one may conclude that this mode of slaking enhances the acceleration. Each wet-dry cycle breaks the sample into progressively smaller blocks, and the sample weathers, in essence, from the "inside out."

Conversely, samples in the decelerated slaking region weather from the "outside in." In this region, five-cycle slake values are less than five times the one-cycle values, and samples lose progressively less material with each wet-dry cycle (as shown on Figure 3). Chip or surface slaking is dominant, and samples slake by sloughing surface layers that are in direct contact with water. Once the weathered and weakened surface layers have slaked (during the first cycle or two), the process decelerates, and less material slakes with each cycle. Although these materials typically have one-cycle slake values higher than samples with accelerated slaking, the five-cycle value is much lower in comparison. Therefore, decelerated samples are "protected" from continued slaking because they slough off outer layers, rather than fracturing internally as do accelerated samples.

As material is weathered, the rate of slaking and the susceptibility to weathering change, even though the mode of slaking may not clearly change. For example, samples Mm21, Mm22, and Mm23 on Figure 4c show a clear progression of weathering. The susceptibility to weathering (five-cycle slake index) increases with each sample, while the rate of slaking (five-cycle to one-cycle ratio) decreases. Sample Mm21 was not observed to slake during the jar slake test, and samples Mm22 and Mm23 showed body or block slaking. This mode of slaking was not expected based on the plot locations on Figure 4c, although it should be noted that it was easier to pry flakes from the more highly weathered sample, Mm23.

Conclusions

Based on the samples evaluated for this study and the discussion presented above, several conclusions may be drawn regarding shale weathering and slaking:

- The jar slake test effectively indicates the mode of slaking, where total degradation represents dispersion slaking, formation of chips represents chip or surface slaking, and formation of fractures represents body or block slaking.
- The mode of slaking may be used to interpret the mode of weathering. Body or block slaking indicates accelerated weathering, where induced fractures rapidly degrade material integrity. Chip or surface slaking indicates decelerated weathering, where surficial layers slough off, and slaking slows down because the exposed surface area decreases as the intact sample reduces in size.
- The slake index test indicates both the mode of slaking and the rate of slaking. These factors can be extrapolated to indicate the mode and rate of weathering in general.
- The one-cycle slake index test corresponds better to slake durability and jar slake than does the five-cycle slake index test. This is probably because the degree of wetting and drying for the one-cycle slake index test is similar to the degree of wetting and drying for slake durability and jar slake tests. The five-cycle slake index test may stress the sample more significantly than the slake durability and jar slake test because the wetting and drying process is repeated so often. The five-cycle test may better represent the long term reaction to weathering in the field.
- Materials do not often show a constant slaking (or weathering) rate, where the five-cycle to one-cycle ratio is near five. Through natural weathering, materials may change modes of slaking and rates of slaking.

Appendix - References

Burwell, E.B. and Moneymaker, B.C., 1950, Geology in Dam Construction: *in* Paige, S., chairman, Application of Geology to Engineering Practice: Berkey Volume: The Geological Society of America, New York, pp. 11-43.

Deo, P., 1972, Shales as Embankment Materials: Ph.D. Thesis, Purdue University, 201 p.

Hudec, P.P., 1982, Statistical Analysis of Shale Durability Factors: Transportation Research Record 873, pp. 28-35.

Mitchell, J.K., 1993, Fundamentals of Soil Behavior: John Wiley & Sons, New York, 437 p.

Morgenstern, N.R. and Eigenbrod, K.D., 1974, Classification of Argillaceous Soils and Rocks: Journal of the Geotechnical Engineering Division, ASCE, vol. 100, no. GT10, pp. 1137-1156.

Moriwaki, Y. and Mitchell, J.K., 1977, The Role of Dispersion in the Slaking of Intact Clay, *in* Sherard, J.L. and Decker, R.S., eds., Dispersive Clays, Related Piping, and Erosion in Geotechnical Projects, ASTM STP 623: American Society for Testing and Materials, pp. 287-302.

Perry, E.F. and Andrews, D.E., 1982, Slaking Modes of Geologic Materials and Their Impacts on Embankment Stability: Transportation Research Record, pp. 22-28.

Santi, P.M., 1995a, Assessing Strength and Durability Properties of Shales, *in* Keefer, D.K. and Ho, C.L., eds., Landslides Under Static and Dynamic Conditions - Analysis, Monitoring, and Mitigation, ASCE Geotechnical Special Publication No. 52, pp. 37-55.

Santi, P.M., 1995b, Classification and Testing of Weak and Weathered Rock Materials: A Model Based on Colorado Shales: unpublished Ph.D. dissertation, Colorado School of Mines, Golden, CO, 286 p.

Walkinshaw, J.L. and Santi, P.M., 1996, Chapter 21 - Shales and Other Degradable Materials, *in* Landslides: Investigation and Mitigation, TRB Special Report: Transportation Research Board, National Research Council, Washington, D.C.

Wood, L.E. and Deo, P., 1975, A Suggested System for Classifying Shale Materials for Embankments: Bulletin of the Association of Engineering Geologists, vol. 12, no. 1, pp. 39-55.

EMBANKMENT DAMS IN THE PIEDMONT/BLUE RIDGE PROVINCE
By Chuck Wilson[1], Member ASCE and Ray Martin[2], Fellow ASCE

ABSTRACT

The paper presents a summary of the geology and typical engineering properties of residual soils applicable to design of earthen embankment dams in the Piedmont and Blue Ridge Province. Several aspects of embankment dam design are addressed. Topics discussed include seepage control, settlement and stability with emphasis on unique aspects associated with the soils and rock of this region.

GEOLOGY

A large portion of the Eastern United States is underlain by similar pre-Cambrian bedrock. The Piedmont/Blue Ridge physiographic province extends from east-central Alabama to northwest New Jersey. The bedrock consists of gneisses, schists and granite intrusives. The residual soils typically classify as silty sands (SM), sandy silts and clayey silts (ML), and lean (CL) and fat clays (CH) . The fines content typically varies from about 30 to 60 percent and mica is usually present. A typical soil profile is more clayey near the surface, where more weathering has occurred, transitioning to more silty and then sandier zones with depth. The granitic soils tend to be more sandy.

The surficial soils usually do not exhibit the structure of the parent rock, but below these soils the structure becomes visible in the saprolitic zone. These soils are effected by stress history due to overburden removal through erosion and by desiccation. They also exhibit residual mineral bonds. Less weathered rock "floaters" or "lenses" are often present in the saprolite. This zone grades transitionally into weathered rock, then rock. Although, the presence of joints, fracture zones and steeply dipping foliation can cause deeper weathering, the rock surface generally represents a subdued reflection of the ground surface. Rock is rarely greater than about 100 feet (30 m) below the ground surface and as elevations increase in the Blue Ridge, rock exists at much shallower depths.

Residual soils can be sampled by conventional undisturbed tube samples for laboratory consolidation testing so long as SPT N-values are below about 10 to 15 and rock fragments are not present. The latter cause damage to tubes and disturbance to the sample. In situ testing, especially by use of the pressuremeter, has been successfully employed where undisturbed tube samples cannot be obtained.

---

[1] Senior Associate, Schnabel Engineering Associates, Inc., Gainesville, Georgia
[2] Principal, Schnabel Engineering Associates, Inc., Ashland, Virginia

The rocks in the Piedmont/Blue Ridge generally trend NE/SW with foliation dips to the SE. The tectonic conditions which created the geologic structure have also resulted in fractures that vary from close to widely spaced. There are numerous faults present that have associated distinct shear zones. The southern portion of the Province and extreme northern area are also located in an area of moderate seismic risk. A word of caution concerning rock. The presence of auger refusal during drilling operations should not be construed as the top of sound rock. Because of the potential for "lenses" or "floaters" of hard rock within both the saprolite and weathered rock, rock coring is required to verify the character and continuity of refusal material.

CLASSIFICATION SYSTEMS

Many classification systems have been proposed for residual soil/rock profiles derived from metamorphic and igneous rocks. The necessity for a classification system arises from the varying engineering properties of the weathering profile. Table 1 summarizes several systems that have been suggested including the one used by the authors' firm.

SEEPAGE CONTROL

Uncontrolled saturation and seepage has been blamed for about one third of all dam failures with piping along conduits as a major problem (1). The 165 foot (50 m) high Bouldin Dam in Alabama failed in 1976 due to piping through backfill along the power house structure. Numerous other failures of lesser magnitude including levees have occurred without loss of life, but with substantial damage to embankments and property. Thus, seepage control is critical in any dam design.

The in situ hydraulic conductivity or permeability of residual soil and rock varies with the zone considered. Sowers (4) suggested the following permeability values.

Materials	Permeability (cm/sec)
Soil	10^{-3} to 10^{-7}
Saprolite	10^{-4} to 10^{-6}
Partially weathered rock	10^{-1} to 10^{-5}
Rock (mass)	10^{-4} to 10^{-6}

The wider variation in the soil category occurs due to the wider variation in the soil classification from silty sand to clay (SM,SC,ML,CL,CH). Saprolite tends to be more narrowly banded into sandy silt and silty sand (SC,SM,ML) and thus the permeabilities are also more narrowly banded. Partially weathered rock may have higher or lower permeabilities than the overlying saprolite depending upon the amount of more nearly rock like material, the fracture condition and the classification of the weathered material which is usually sandy in texture. Permeability in metamorphic rock can be anisotropic with the permeability parallel to fracturing trends up to ten times or more greater than the permeability perpendicular to the fracture trends.

Because of the potential for variations in the permeability of both foundation and embankment materials, in situ and laboratory permeability testing is usually a prudent

TABLE 1 - ENGINEERING CLASSIFICATION SYSTEMS OF WEATHERING PROFILES FOR IGNEOUS AND METAMORPHIC ROCKS (EXTRACTED IN PART FROM REFERENCE 4)

SOWERS	DEERE AND PATTON		LAW/MARTA	AUTHORS' FIRM
IGNEOUS AND METAMORPHIC ROCK	IGNEOUS AND METAMORPHIC ROCK		IGNEOUS AND METAMORPHIC ROCK	IGNEOUS AND METAMORPHIC ROCK
SOIL $N = 5 - 50$	I RESIDUAL SOIL	IA A HORIZON	UPPER HORIZON NO RESIDUAL STRUCTURE	RESIDUAL SOIL $N < 60$
		IB B HORIZON		
SAPROLITE $N = 5 - 50$		IC C HORIZON (SAPROLITE)	SAPROLITE	
PARTIALLY WEATHERED ROCK ALTERNATE HARD & SOFT $N > 50$	II WEATHERED ROCK	IIA TRANSITION FROM RESIDUAL SOIL TO PARTLY WEATHERED ROCK	PARTIALLY WEATHERED ROCK $N > 100$ CORE RECOVERY<50%	DISINTEGRATED ROCK $N \geq 60$
		IIB PARTLY WEATHERED ROCK		
			ROCK CORE RECOVERY>50% RQD < 50 %	ROCK $N \geq 100/2$ in CORE FOR CONFIRMATION
ROCK RQD > 75%	III UNWEATHERED ROCK RQD > 75%		SOUND ROCK RQD > 50% CORE RECOVERY>85%	

RQD = ROCK QUALITY DESIGNATION
N = STANDARD PENETRATION TEST N VALUE

1 in = 0.025 m

course of action in the design of new dams with a height greater than about 35 to 45 feet (11 to 14 m). In situ falling head permeability testing is typically performed on residual soil and partially weathered rock. In situ permeability testing with double packers is most often utilized to assess the seepage potential of intact rock. Permeability testing of remolded samples of potential embankment fill soils is performed in the laboratory. Such remolded soils are usually compacted to at least 95 percent of the soils maximum dry density and at a moisture content consistent with design requirements as determined by the Standard Proctor compaction test, ASTD-D698. Higher compaction levels are sometimes tested when a greater reduction of seepage flow through the embankment is required.

Foundation Seepage Control

The need for cutoff trenches, foundation grouting, downstream relief wells and other methods of foundation treatments to reduce and control seepage through the foundation and abutments of a dam depends on the permeability of the various layers and the reservoir head. Dams of over 100 feet (30 m) in height have been designed in this geologic setting without grout curtains and in some cases without cutoff trenches, except through alluvial deposits adjacent to streams. These designs are appropriate because of the low permeability of the foundation rock and overlying soils.

If the more clayey residual soils at the ground surface are relatively impervious and deep, a cut off trench may be of little benefit except to cutoff loose alluvial soils if they are left in-place in the valley bottom. This natural condition is similar to installing an upstream impervious blanket. On the other hand, when more pervious residual soils occur from the ground surface, a cutoff trench consisting of compacted fine-grained soils may be required. Under very high head conditions, of say several hundred feet, a cut off to rock would likely be required. Under low head conditions, of say less than 20 feet (6m), a shallow cutoff to a depth of say 5 feet (1.5 m) may be required to eliminate animal borrows. A downstream filter layer may also be appropriate adjacent to the cutoff trench if fractures are significant in the disintegrated rock or rock.

The need for surface treatment of rock at the bottom of a cutoff trench or rock grouting must also be considered. Surface treatment usually consists of cleaning out any soil filled rock fractures and backfilling with dental concrete. Overhangs of rock should also be cut back or backfilled with dental concrete. Slush grout is often applied to the surface of the rock to fill very thin fractures typical of the regions rock. This is normally done after cleaning with high pressure air or water. Grouting is usually performed to fill fractures at depth and may only be performed in selected areas along the axis of the dam based on test boring data and geologic mapping of the cutoff trench. Void filling is usually not a concern in the region because of the type of rock. The general absence of carbonate rock reduces the concern for solution cavities. However, some marble is locally present within the Triassic Basins of the Piedmont and these rocks can be solutioned. The number of grout lines and depth of grouting is dependent on the desired reduction of seepage. Normally grouting to depths of greater than 100 feet (30 m) is counterproductive unless reservoir heads are very high. Single-line grout curtains are the norm in this geology. Initial spacing may be on the order of 10 to 20 feet (3 to 6 m)with split spacing in areas of significant grout takes. Grouting will not eliminate seepage but can reduce the mass permeability by about one half to one order of magnitude.

Another method of controlling seepage and reducing uplift pressures in the foundation is through the use of relief wells. These wells are usually designed to be located in the down stream one third of the embankment foot print. They may be used in conjunction with a grout curtain for very porous foundation conditions or in lieu of a grout curtain when under seepage is expected to be small. Well depth is dependent upon a seepage analysis. Wells can be extended into residual soil, partially weathered rock or rock. Screen and filter design are imperative to provide wells which will function properly over the life of the dam. Clean-outs should be provide to allow flushing should problems occur. These wells are usually discharged independently of the internal drainage system of the dam to allow identification of flow volume from different sources in the dam for discharge. In some cases, a downstream blanket drain or a series of trench drains placed perpendicular to the dam axis can serve the same purpose as relief wells.

Typically, soft alluvial soils which blanket floodplain areas are undercut within the footprint of new dams. There have been several instances where the authors have recommended the construction of an earthen embankment dam on an extremely soft alluvial foundation. In such cases, the presence of deep and laterally extensive alluvial material within the dam footprint would have resulted in excessively high construction costs for its removal. The alluvium was left in place and seepage control through the foundation was achieved by the use of bentonite slurry vibrated beam cutoff wall (5). The bentonite cutoff was extended through the alluvium and was "socketed into" residual material for at least two feet. Such a design was successful because of the relatively low height of the embankment (less than on 50 feet (15 m)). Additional measures, such as flattened slopes and additional internal drainage structures were necessary.

Embankment Seepage Control

In general, the residual soils of these two physiographic provinces can be used in embankment dam construction. Most embankments in the region are constructed with a homogeneous cross section or with a modified homogeneous cross section where the less permeable materials are placed in the central section of the dam and the more permeable materials are placed in the outer portions of the dam. Some dams have been designed with a more formalized zoning based on the same approach. The core material is usually obtained from the upper portion of the soil profile which contains higher clay content soils. Core material can also be obtained from colluvium deposits that blanket swales and floodplain areas. It is common practice to increase the moisture content of core material above optimum moisture content to reduce permeability and to increase plasticity. Greater compactive effort can also be used to reduce permeability.

In the design of a central core, the authors have found the following criteria obtained from Sowers, to be a good guide:

$$b = 20 + 0.1 \, \Delta h \text{ (for clay cores)} \tag{1}$$

$$b = 20 + 0.3 \, \Delta h \text{ (for silt cores)} \tag{2}$$

where b is the core thickness at any elevation, and Δh is the head difference at that point.

Filters and drains are used for internal seepage control within a dam and to enhance stability. As a general rule, dams constructed of residual soils and with a height greater than 20 feet (6 m) should be constructed with an internal drain. Drainage blankets are commonly used for the condition of a foundation and embankment of similar permeability. Thus, the drainage blanket collects ground water seeping up from the foundation and down through the dam and maintains the seepage line below the down stream surface. Drainage blankets typically extend from the downstream toe under the dam over about 25 to 35 percent of the base. They also usually extend up each abutment to just below the normal pool level. The blankets should be designed with natural filters and drainage layers. Geotextiles should only be used when they can be repaired such as in a toe drain. Narrower strip drains can also be used in lieu of blanket drains if the seepage analysis indicates adequate control of the phreatic surface and sufficient pressure relief at the toe of the dam to avoid excessive gradients.

Another form of internal drainage system is the chimney drain. These inclined or vertical drains are often employed in dams with a central core and are located just downstream of the core . They are useful when there is the potential for a seismic loading condition which may cause cracking of the core. Chimney drains are also beneficial when the embankment is transitioning across the different foundation materials, such as a soil-rock or soil-alluvium interface. A word of caution on drain construction. For dams in this region the presence of abutment seepage occurring several years after construction has often been a problem. The authors have examined numerous engineered dams where additional seepage measures had to be implemented along the abutment-dam interface after the dam was constructed and the lake impounded to normal pool. Although drains can be constructed to provide sufficient seepage control within the foundation and embankment, seepage within the abutments can sometimes be unpredictable based on the particular site geology.

Seepage control along structures penetrating dams is also a concern (2). These structures, whether a pipe or a cast-in-place structure, represent a discontinuity in the dam and provides a preferential path for seepage. Residuals soils are especially susceptible to erosion due to the silty texture of the soils and care must be taken to adequately protect against this type of problem. The possibility of achieving less than the required quality compacted backfill adjacent to a conduit can also increase piping potential. As a further safeguard in design, fill soils should be checked to assess their dispersion potential.

All pipes penetrating embankments should be designed with concrete cradles except for very low head conditions on the order of about 6 feet (2 m). Cradles not only provide pipe support, but also provide a better condition for placement and compaction of backfill upstream of the drainage layer. Cradles provide an excellent means of seepage control along conduits because of difficulties in the placement of compacted fill soils beneath the springline of any pipe. Cradles should be articulated on soft foundations to allow for rotational settlement under the dam and should match pipe joints. As a general rule, if the pipe and cradle are to be placed on a yielding foundation, the pipe should be designed with a camber. The height of camber is based on the anticipated settlement of the foundation beneath the load of the planned embankment.

Two methods can be considered for construction of cradles. There is some disagreement among "dam engineers" as to the benefits and drawbacks of the two methods. One method involves forming the sides of the cradle and then placing compacted fill against the sides of the cradle. The sides are battered inward to provide for better compaction of the soil against the cradle. Another method is to place the conduit in a trench excavated in either residual soil or compacted backfill. The pipe is assembled in the trench. The concrete for the cradle is then placed beneath the springline of the conduit, with the sidewalls of the excavation serving as the formwork. Because concrete may shrink away from the vertical sidewalls of the excavation, strips of bentonite can be placed perpendicular to the direction of pipe flow at several locations along the trench.

Conduits must be designed to be watertight under both the maximum internal hydraulic pressure and the maximum external embankment load. Joints should be designed to allow for rotation and extension due to settlement of the dam foundation and should incorporate watertight gaskets. The length of the joint must allow for the estimated extension due to settlement and should include a factor of safety of 0.5 in (1.27cm).

Proper treatment of rock surfaces must be provided similar to that presented for cutoff trenches. When pipes or other conduits extend below the surface of rock, the excavation should be backfilled to the surface of rock with lean concrete, or alternatively, the structure may be cast against the rock. Concrete should be formed, as described above, to achieve good backfill compaction if the surface of the rock is below the spring line of the pipe. If at all possible, a conduit should be placed entirely on either soil or rock. If rock is near the surface, consideration should be given to extending the cradle to rock.

Prior to about 1970, anti-seepage collars were included in the designs of most dams as a seepage control measure. Today properly designed natural graded filters are used to prevent the movement of soil fines and to reduce the piping potential. Filter diaphragms are recommended to be used to prevent piping along conduits. The diaphragm should be located downstream of the cutoff trench or core zone or downstream of the centerline of a homogeneous dam. The drainage layer within the diaphragm should be surrounded by an appropriate filter and should be connected to the internal drainage system of the dam. The filter diaphragm may be eliminated if a chimney drain is provided. The conduit or structure may also be completely encircled by the foundation drainage system downstream of the core when supported on an earth foundation. On rock foundations the drainage layer should extend around the conduit above the surface of the rock. For homogeneous dams on a soil foundation the drainage layer should extend along the downstream one quarter to one third of the conduit or structure.

EMBANKMENT/FOUNDATION SETTLEMENT

Soil compressibility varies with classification. For the stresses normally imposed by a dam in this geologic setting rock compressibility is not an issue of concern. Settlement of the embankment soils under their own weight can be significant, but foundation settlement may be less than anticipated because of the stiffness imparted by the relic structure of the rock. Sowers (4) reported void ratios for undisturbed soils range as follows:

Material	Void Ratio
Soil - sandy	0.6 to 1.0
Soil - clayey	0.4 to 0.8
Saprolite	0.7 to 3.0
Partially weathered rock	0.1 to 0.5
Rock	Less than or equal to 0.02

The void ratios in saprolite can be very high if mica is present. These soils typically exhibit a "preconsolidation effect" unrelated to depth or preconsolidation stress. The preconsolidated effect results from residual mineral bonds between partially or unweathered grains. Typical values range from 0.5 to 2.5 tons/sf (47.9 to 239.4 kN/m^2.)

These factors produce higher strength and lower compressibility than would be expected. Sowers also reported a correlation between void ratio, e, and compression index, C_c, as follows:

$$C_c = 0.75 (e - 0.55)$$

The pressuremeter has also been used to estimate residual soil settlement. Martin (3) proposed a relationship as follows:

$$E_{pm} = \log^{-1} (1.176 + 0.704 \log N)$$

where E_{pm} is the pressuremeter modulus, and N is the Standard Penetration Test N-value.

Embankment settlements must be more carefully considered, especially for larger dams. Residual soils and saprolite typically have lower compacted dry densities and are more compressible than similarly classified sedimentary soils due to the presence of mica. The authors have found typical dry density values range from about 90 to 110 pcf (14.1 to 17.3 kN/m^3) and optimum moisture contents range from about 15 to 25 percent. One method to reduce compressibility is to use Modified Proctor ASTM D1557, as the compaction standard. However, it is much harder for a contractor to achieve 95 percent of this standard than say 95 percent of Standard Proctor since the energy is about four times a great. This will cause the cost of the dam to increase. The use of this higher density standard must be considered carefully to make sure it is warranted.

In many cases, the construction duration for an earthen embankment dam in either the Piedmont or Blue Ridge is of such a length, that much of the primary consolidation within the fill has taken place before the dam is completed. However, the authors typically recommend overbuilding an embankment crest by several feet to compensate for any long term secondary compression of either the fill or foundation soils.

EMBANKMENT SLOPE STABILITY

The side slopes for smaller embankment dams are often established based on correlations with material type available for construction of the embankment. This approach is used in lieu of performing shear strength testing and stability analyses. The authors' experience indicates that in most cases, Piedmont and Blue Ridge soils, when compacted to at least 95 percent of the Standard Proctor maximum dry density, will provide for an adequately stable embankment slope when placed at a grade of 3H:1V on a reasonably firm/stiff residual soil foundation.

For larger dams testing and stability analyses are required. Total and effective strength parameters are used in most dam designs. These parameters are usually obtained from consolidated undrained (CU) triaxial tests with pore pressure measurements. Sowers (4) complied triaxial test data obtained for design of the Atlanta subway system and plotted a failure envelope trend line for which 1/3 of the data were below the trend line. These tests were performed on samples of both saturated residual soils derived from both gneisses and schists. The authors have also tested many samples of similar residual soils derived from gneisses and schists. These tests have been performed on samples compacted to about 95% of Standard Proctor maximum dry density and generally at moisture contents slightly above optimum moisture. Results are shown in Table 2. Residual soils can exhibit a true effective cohesion (shear strength at zero confinement) due to residual mineral bonds.

TABLE 2 - CONSOLIDATED UNDRAINED (CU) TRIAXIAL DATA

PARAMETERS	SOWERS (4)		AUTHORS	
	TOTAL	EFFECTIVE	TOTAL	EFFECTIVE
ANGLE OF INTERNAL FRICTION, ϕ	23°	32°	10° to 23°	25° to 34°
COHESION, c	270 psf	0	430 to 1500 pcf	0 to 460 psf
1 psf = 0.049 kN/m^2				

In evaluation of embankment dam slopes constructed of residual soils end-of-construction, steady-state seepage, rapid drawdown of the upstream slope, and steady-state seepage with seismic loading are usually analyzed. In assessing slope stability, soil parameters reflecting the stress-state of the soil for each case are utilized in the analysis. With few exceptions, the most critical case is steady state seepage. This case is even more critical when seismic risk is higher. Typically, a pseudo-static analysis is performed in which the embankment is subjected to horizontal and, in some cases, vertical, seismically induced loadings. There is much debate in the regional geotechnical engineering community as to the appropriateness of this analysis. The debate centers on whether or not a deformation analysis is more appropriate for areas which could be expected to experience higher ground accelerations.

Another factor concerning embankment stability involves the problem of soil erosion. In the construction of dams in the region, much, if not all, of the more clayey cohesive soils are used in either a defined "clay" core or within central portion of the embankment. As a result, both upstream and downstream shells, or outer portions of the embankment, are constructed of the more sandy/silty non-cohesive material. Such soils are susceptible to erosion, even if a good stand of grass has been established. On the upstream slope at the normal pool level, erosion protection for wave action is usually provided by a rip-rapped berm. The use of vinyl sheet piles for this purpose are being tested on a middle Georgia dam at this time. On downstream slopes of dams in excess of 50 feet (15 m) high, intermediate berms are usually incorporated into slopes to divert surface flow along slope to either abutment and to reduce the potential for erosion.

SUMMARY

The following key issues were discussed concerning design of dams in the Piedmont/Blue Ridge.

- The weathering profiles derived from the gneiss, schist and granite rocks underlying much of the region typically consists of more clayey soils near the surface with more sandy soils at depth. These lower soils often contain rock "floaters".

- The rock and residual soils of the region are generally suitable for support of embankment dams.

- Foundation seepage control can often be accomplished with cutoff trenches, relief wells and limited grouting of only highly fractured zones. Some dams have been designed without cutoff trenches due to low permeability surficial soils. Typical permeability data are presented herein for rock and in situ residual soil.

- Embankments dams typically have been designed with central zones consisting of more clayey soils with chimney and blanket drains for internal seepage control. Only graded filters should be used on the interior of a dam.

- Pipe and structural penetrations of dams must pass through a chimney drain or filter diaphragm for adequate seepage control. Pipes should be supported on cradles which are articulated for yielding foundations.

- Three horizontal to one vertical side slopes are commonly employed in designing embankment dams with these soils. Data are provided herein concerning typical shear strength and compressibility values of residual soils for stability and settlement analyses.

ACKNOWLEDGMENTS

The authors wish to thank Schnabel Engineering Associates, Inc., for providing funding for preparation of the paper.

REFERENCES

1. Martin, R. E., "Embankment Dam Safety", Proceedings, Geotechnical Practice in Dam Rehabilitation, ASCE, NCSU, Raleigh, North Carolina, 1993.

2. McGill, D. M., and Martin, R. E., "Conduits Through Embankment Dams" ASDSO Mid Atlantic Regional Dam Safety Seminar, Mendenhall, Pennsylvania, 1995.

3. Martin, R.E., ""Settlement of Residual Soils", Proceedings, Geotechnical Engineering Symposium, Spring Convention, ASCE, Atlantic City, New Jersey, 1987.

4. Sowers, G. F., "Residual Soils of the Piedmont and Blue Ridge", Transportation Research Board, Research Record No. 919, 1983, pp. 10-16.

5. Wilson, C.B., "Construction of Earthen Dams on Very Soft Foundations", Southeastern States Dam Safety Conference, Williamsburg, Virginia, 1990.

In Situ Measurement of Rockfill Properties

Anne Eckert Clift, Ph.D.[1]

Abstract

In situ testing was performed at three rockfills in western Colorado. These rockfills vary in type of construction material (sandstone or marlstone), thickness (maximum thicknesses from 150 to 315 feet), and compaction conditions (compacted or dumped). Pressuremeter, point load, and large-scale in situ bulk density tests were performed. Data analysis showed that in situ deformation moduli from the sandstone fill were greater than those from the marlstone fills. Moduli and limit pressures were not found to vary with depth (test depths ranged from 3.5 to 48 feet), compactive effort, compressive strengths (estimated from point load), dry unit weight, or grain size. It seems that the volume of material tested by the pressuremeter is too small in comparison to the large variation of particle sizes in rockfill. As a result, the pressuremeter may be limited in its use in rockfills.

Introduction

Deep rockfill is becoming increasingly important as foundation material for large industrial complexes, residential developments, and highways in mountainous terrains. Two factors which must be considered in the geotechnical design of these structures include settlement behavior and slope stability. However, measurement of rockfill engineering properties is difficult. Particle sizes can range from clay size up to boulders, six feet or larger in diameter. It is therefore not only a challenge to test representative samples in a laboratory environment, but also difficult to bring undisturbed samples to the laboratory. The alternative, in situ

---

[1] Consultant, RR1 Box 6, Warren, VT 05674

testing, is expensive and time-consuming and as a result, has seldom been performed.

As part of an analysis of rockfill settlement behavior, several in situ methods, including pressuremeter, large-scale density, and point load tests, were used to estimate engineering properties at three rockfills at the Exxon Colony Shale Oil Project site in western Colorado (Clift, 1994). The rockfills vary in type of construction material (sandstone or marlstone), thickness, and compaction conditions (compacted or dumped).

Description of Rockfills

Data were collected at the Mine Bench Fill, the Primary Crusher Pit, and the South Valley Fill located at the Exxon Colony Shale Oil Project site about 45 miles north-northeast of Grand Junction, Colorado. The fills are located in the valley of the Middle Fork of Parachute Creek.

The Mine Bench Fill, shown in Figure 1, is composed of oil shale and marlstone from the Eocene Green River Formation. The fill is 700 to 800 feet wide at the top and is up to 315 feet thick. It was randomly dumped in place in 6- to 8-foot thick lifts. Compaction of this rockfill occurred only as the result of truck traffic over the site. Grain size data and typical gradation curves are shown on Table 1 and Figure 2 respectively.

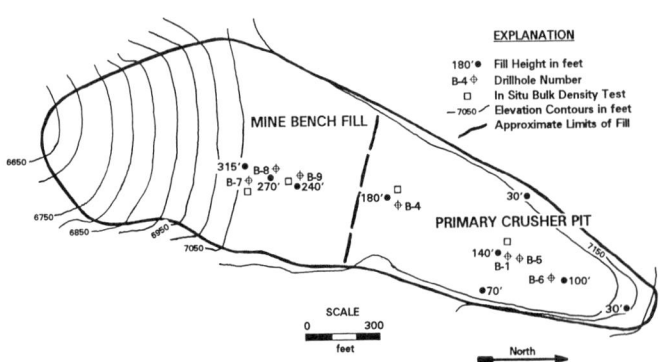

Figure 1. Map of Mine Bench Fill and Primary Crusher Pit

Table 1. Grain Size Data

Fill Name	%Passing #4 Sieve	%Passing #200 Sieve	Coeff. Uniform.	Coeff. Curv.	Atterberg Limits Liq.Lim.	Plas.Ind.
MBF	12.2	1.11	23.3	2.30	22.7	2.0
MBF	14.2	1.21	20.0	1.57	26.3	0.5
PCP	38.3	4.02	37.1	1.37	25.7	5.2
PCP	20.3	3.37	36.7	2.15	23.3	3.2
SVF	41.1	3.06	110.0	0.15	Nonplastic	
SVF	41.8	3.27	90.0	0.14	Nonplastic	

Note: MBF - Mine Bench Fill; PCP - Primary Crusher Pit; SVF - South Valley Fill

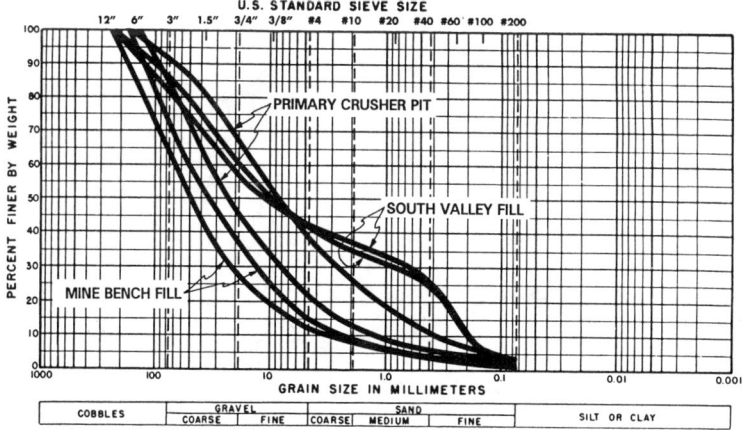

Figure 2. Gradation Curves

The Primary Crusher Pit, shown in Figure 1, lies upstream of the Mine Bench Fill and is also composed of oil shale and marlstone of the Eocene Green River Formation. This fill ranges from 150 to 650 feet wide and is up to 150 feet thick. It was compacted in maximum two-foot lift thicknesses with at least six passes of a smooth drum vibratory compactor (Matheson and Parent, 1989). Grain size data and typical gradation curves are shown on Table 1 and Figure 2 respectively.

The South Valley Fill, shown in Figure 3, is composed of sandstone, siltstone and marlstone of the Eocene Uinta Formation. The fill is approximately 450 feet wide and up to 173 feet thick. The final specifications included a maximum lift thickness of 2 feet and at least six passes of a smooth drum vibratory compactor (Matheson and Parent, 1989). Grain size data and typical gradation curves are shown on Table 1 and Figure 2 respectively.

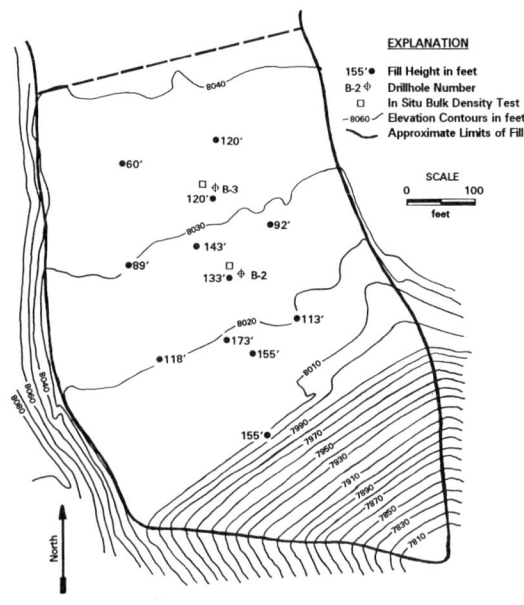

Figure 3. Map of South Valley Fill

Subsurface Investigation

The subsurface investigation of the rockfills included coring into the rockfills, pressuremeter testing, point load testing on selected core samples, and large-scale in situ density tests. The reader is directed to Clift (1994) for additional details of the subsurface investigation.

ROCKFILL PROPERTIES MEASUREMENT

A total of 200 feet was cored in the three rockfills. All holes were cored with mud using a wireline NX core system. Circulation was lost in all holes. Overall core recovery ranged from 32 to 56% in the Mine Bench Fill, 33 to 48% in the Primary Crusher Pit, and 59 to 72% in the South Valley Fill. These low core recoveries reflect the unconsolidated nature and the void spaces in the rockfill. Drillhole locations are shown in Figures 1 and 3.

A Menard G-Am Pressuremeter was used in this investigation. Of the 43 pressuremeter tests attempted, 31 were successful. Tests were unsuccessful due primarily to ruptured pressuremeter membranes.

Point load tests were performed on the larger pieces of core to obtain an estimate of intact rock compressive strength (International Society for Rock Mechanics, 1985). Mean compressive strengths were estimated to be 13,800 psi in the Mine Bench Fill, 15,100 psi in the Primary Crusher Pit, and 6,300 psi in the South Valley Fill. Average results for each hole are shown on Table 2.

Two in situ large-scale density tests were completed on each of the three rockfills. For each test, about 6 to 10 cubic feet of material was removed from the interval between approximately three to five feet below the ground surface. The average dry unit weight of the rockfill was 94 pcf in the Mine Bench Fill, 124 pcf in the Primary Crusher Pit, and 97 pcf in the South Valley Fill. Locations of the tests are shown in Figures 1 and 3. Test results are shown on Table 2.

Table 2. Point Load and Bulk Density Test Results

Hole No.	Fill	Depth (ft)	Avg.Pt.Load Comp.Str. (psi)	Dry Unit Weight (pcf)
B-1	Primary Crusher Pit	49.0	14,700	-
B-2	South Valley Fill	51.0	7,900	92
B-3	South Valley Fill	25.0	3,600	103
B-4	Primary Crusher Pit	12.5	17,300	128
B-5	Primary Crusher Pit	12.5	13,000	120
B-6	Primary Crusher Pit	12.5	14,500	-
B-7	Mine Bench Fill	12.5	14,500	103
B-8	Mine Bench Fill	12.5	11,400	-
B-9	Mine Bench Fill	12.5	14,900	86

Results and Analysis of Pressuremeter Testing

An ideal pressuremeter curve is shown in Figure 4. For each depth tested, a plot of pressure versus volume is generated. In situ deformation modulus is calculated using the pseudo-elastic part of the pressure-volume curve (see Figure 4). Analysis of pressuremeter data was performed according to the methods discussed in Baguelin et al. (1978). The calculated deformation modulus for each test depth is listed on Table 3.

The pressuremeter tests completed did not reach limit pressure (defined in Figure 4). It is questionable in many cases whether the tests reached creep pressure (defined in Figure 4). Therefore, limit pressures were estimated using the average values estimated from the manual extrapolation and upside-down curve methods as described in Baguelin et al., 1978 (pp. 425-430). The resulting limit pressure values are shown on Table 3.

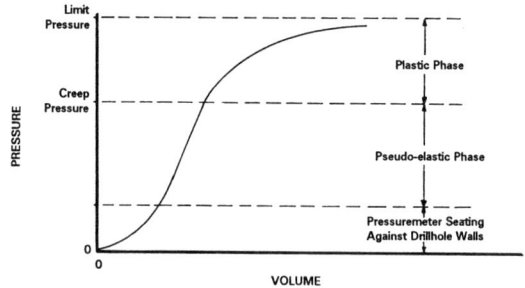

Figure 4. Ideal Pressuremeter Curve

The relationship between the pressuremeter moduli and limit pressures is shown in Figure 5. Results from the South Valley Fill (predominantly sandstone) are shown as square points. Results from the Primary Crusher Pit and Mine Bench Fill (both predominantly marlstone) are shown as diamonds. The circled points indicate those tests in which the creep pressure is readily apparent on the pressure-volume plots. The pressure-volume plots of the other tests were straighter. Therefore, there is less confidence in the creep pressures identified on those plots and the limit pressures estimated using those plots.

Table 3. Results of Pressuremeter Testing

HOLE NO.	FILL*	DEPTH OF TEST (ft)	IN SITU. DEF.MOD. (psi)	LIMIT PRESSURE (psi)
B-2	SVF	4	4,700	170
		6.5	3,500	280
		9	4,500	470
		11.5	3,100	310
		14	6,500	560
		16.5	2,400	360
		19	13,800	1,040
		21.5	16,300	750
		24	4,100	390
		28.5	3,200	330
		31	12,300	770
		33.5	4,000	340
		36	8,200	470
		43	4,800	630
		45.5	6,500	810
		48	8,100	780
B-3	SVF	4	5,200	470
		9	9,500	650
		14	4,800	310
		19	3,000	330
		24	1,500	150
B-4	PCP	4	2,700	350
		9	800	130
B-5	PCP	7	4,500	510
B-6	PCP	5.5	1,400	190
		10.5	2,400	270
B-7	MBF	5.5	4,700	420
		10.5	6,200	410
B-8	MBF	5	3,100	380
B-9	MBF	3.5	2,400	200
		7	2,000	250

* FILL: SVF = South Valley Fill (compacted sandstone)
PCP = Primary Crusher Pit (compacted marlstone)
MBF = Mine Bench Fill (uncompacted marlstone)

In general, Figure 5 shows that as deformation modulus increases, so does limit pressure. A regression line through the five points in which creep pressure has been identified with some confidence has a lower slope than a regression line through the points in which creep pressure may not have been reached. This suggests that either the limit pressures have been overestimated for those points in which creep pressure may not have been reached or the deformation moduli have been underestimated. However, because deformation modulus is calculated using data collected during the pseudo-elastic phase of deformation which occurs prior to creep pressure, it would seem that the estimates of limit pressure are high for the points which may not have reached creep pressure.

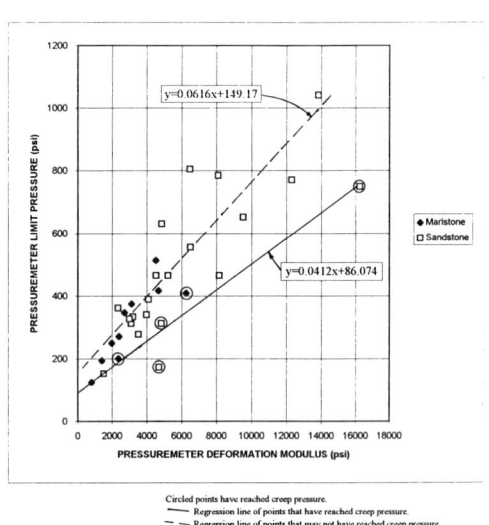

Circled points have reached creep pressure.
—— Regression line of points that have reached creep pressure.
— — Regression line of points that may not have reached creep pressure

Figure 5. Relationship Between Deformation Moduli and Limit Pressures

The moduli and limit pressure values were evaluated to identify variations with depth in the drillholes. Because no more than two tests were conducted in each hole at the Primary Crusher Pit and the Mine Bench Fill, the only data available to investigate pressuremeter test variation with depth are from the South Valley Fill.

Sixteen tests were performed in hole B-2 and five tests were performed in hole B-3 on the South Valley Fill. Both pressuremeter moduli and limit pressures do not appear to vary regularly with depth in these holes.

The moduli and limit pressures in different rockfills were compared. Average modulus and limit pressure values for the predominantly sandstone South Valley Fill (6200 and 490 psi respectively) are higher than the average values for the predominantly marlstone Primary Crusher Pit (2400 and 290 psi respectively) and Mine Bench Fill (3700 and 330 psi respectively). Data from tests less than 10.5 feet deep in the South Valley Fill were compared with data from tests of similar depths in the marlstone fills (Primary Crusher Pit and Mine Bench Fill). The mean modulus and limit pressure for the five shallow tests in the South Valley Fill are 5500 and 410 psi, respectively. The mean modulus and limit pressure for the ten shallow tests in the marlstone fills are 3000 and 310 psi, respectively. The mean modulus of the sandstone fill was determined to be statistically higher than that of the marlstone fills at the 0.05 level of significance. However, the mean limit pressures of the sandstone and marlstone fills were not found to be statistically different at the 0.05 level of significance.

The mean deformation modulus of the compacted fills (South Valley Fill and Primary Crusher Pit) was statistically tested for equality with the mean deformation modulus of the dumped fill (Mine Bench Fill). At the 0.05 level of significance, these means were found to be equal. This suggests that, for the materials tested, deformation modulus is not significantly influenced by compactive effort.

Discussion

An attempt was made to correlate results from the different in situ tests. Pressuremeter data were investigated with respect to the compressive strengths of the rock in the fills estimated from the point load tests (as reported on Table 2). Based on plots constructed of the mean pressuremeter deformation moduli and limit pressures calculated for each hole versus average estimated compressive strength for each hole, moduli and limit pressures were not found to vary with differing compressive strengths.

Pressuremeter data were also investigated with respect to the dry unit weights calculated from the bulk density tests and grain size data. Relationships from plots constructed using the dry unit weights (as reported on Table 2) and the mean deformation moduli and limit pressures calculated for each hole were inconclusive. Correlations between the pressuremeter data and the grain size data were also inconclusive.

The results of pressuremeter testing show a large variation in deformation moduli and limit pressures, even in the same type of rockfill. This variation is probably due to a combination of factors. Because the rockfill is composed of such a wide range of particle sizes and because the pressuremeter only tests a small portion of rockfill at any one time, it is probable that the sample size in this study is too small to make any conclusions about the variation of pressuremeter moduli with depth or with differing compactive effort.

Because of this variability in results, the applicability of the pressuremeter in rockfill may be limited. For example, the pressuremeter is used in soil successfully to evaluate the effectiveness of ground improvement techniques, such as dynamic compaction. However, because of the wide variation seen in this study between compacted and uncompacted rockfills, the pressuremeter may not be appropriate for this purpose in rockfills.

Conclusions

In situ testing was conducted at three rockfills in western Colorado. The rockfills vary in construction material (sandstone or marlstone), thickness (maximum thicknesses from 150 to 315 feet), and compaction conditions (compacted or dumped). Pressuremeter, point load tests, and in situ large-scale density tests were performed at the fills.

Pressuremeter tests were performed at different depths (ranging from 3.5 to 48 feet) in compacted sandstone, compacted marlstone, and uncompacted marlstone fills. Deformation moduli and limit pressures of the sandstone fill ranged from 1,500 to 16,300 psi and 150 to 1,040 psi respectively. Deformation moduli and limit pressures of the marlstone fills ranged from 800 to 6,200 psi and 130 to 510 psi respectively. Moduli and limit pressures were not found to vary with depth of test up to 12 feet deep or be influenced by compactive effort.

Mean compressive strengths were estimated to be 14,300 psi in the marlstone fills and 6,300 psi in the sandstone fill using point load test data on core samples. Relationships between compressive strength and pressuremeter deformation moduli and limit pressures were investigated, however, pressuremeter data were not found to vary with differing compressive strengths.

Large-scale bulk density tests were conducted at each rockfill. The average dry unit weight of the uncompacted and compacted marlstone fills and the compacted sandstone fill were 94 pcf, 124 pcf and 97 pcf respectively. An attempt was made to correlate pressuremeter data with rockfill dry unit weight and grain size data, but relationships were inconclusive.

As a result of the lack of correlation of pressuremeter data to depth, compactive effort, compressive strength, and dry unit weight, the pressuremeter may be limited in its use in rockfills. The large range of particle sizes in rockfill necessitate large test samples, but because the pressuremeter tests only a small volume of rockfill, the results vary widely.

Conversion Factors

1 foot (ft) = 0.348 meters
1 mile (mi) = 1.61 kilometers
1 pound per square inch (psi) = 6.895 kiloNewtons per square meter
1 pound per cubic inch (pcf) = 157 Newtons per cubic meter

Acknowledgments

The author wishes to thank Exxon Coal and Minerals, who provided funding and access to the site.

References

Baguelin, F., Jezequel, J.F., and Shields, D.H., 1978. The Pressuremeter and Foundation Engineering: Series on Rock and Soil Mechanics, Vol. 2, No. 4 (1974/77), Trans Tech Publications, Clausthal, Germany.

Clift, Anne Eckert, 1994. In Situ Properties and Settlement Characteristics of Rockfill. Ph.D. Dissertation T-4375, Colorado School of Mines, Golden, Colorado.

International Society for Rock Mechanics, 1985. Suggested Method for Determining Point Load Strength. International J. Rock Mech. Min. Sci. & Geomech. Abstr. Vol. 22, No. 2, pp. 53-60.

Matheson, G.M. and Parent, W.F., 1989. Construction and Performance of Two Large Rockfill Embankments. ASCE Journal of Geotechnical Engineering, V. 115, No. 12, December, pp. 1699-1716.

Rethinking Foundation Design in Karst Residuum

Raymond A. DeStephen[1], M. ASCE
Steven E. Conner[2], M. ASCE

Abstract

Residual soils formed from carbonate rocks are derived in an entirely different manner than those derived from the in-place physical and chemical weathering associated with all other rock types. This unique weathering, actually a solutioning process, leaves the geotechnical engineer with a host of difficult foundation challenges which must be resolved in order to create a successful project. Chief among these are: a) random zones of very soft soils b) highly variable rock surfaces with slots and pinnacles, and c) solution voids and cavities. All of these aspects of the karst residual profile have major cost and feasibility implications. More importantly, we cannot apply traditional soil mechanics analyses to these features with any certainty. No longer can the geotechnical community ignore the uncertainty inherent with these karstic features, without defining the level of risk of future subsidence for the owner. This can be accomplished by assessing both the risks associated with the site's geology, and those associated with changes to site conditions and the state of stress caused by new construction. These must be conveyed to the owner, along with the alternative solutions that would limit or reduce risks, such that the owner can make an informed decision on how to proceed, if at all.

Karst Residuum

Most residual soil profiles, whether from granite, schist, gneiss, sandstone or shale, are all similar in that the weathering is more pronounced near the surface than at depth. Because the physical and chemical weathering process takes place from the ground surface downward, the soils typically become more competent and "rock-like" with depth. For the geotechnical engineer, this predictability is heavily counted upon when investigating sites and estimating foundation response. This is not so for residual soils formed from carbonate rocks.

[1] Principal, Schnabel Engineering Associates, One W. Cary St., Richmond, Virginia 23220
[2] Sr. Associate, Schnabel Engineering Associates, 2601 S. Main St., Blacksburg, Virginia 24060

Karst residuum is the result of dissolution of calcium and magnesium carbonate rocks. Reference here will be made to the Cambro-Ordovician age limestones and dolomites associated with the Appalachian Ridge and Valley located in the Eastern United States from New York to Alabama. The residual soils in these karstic areas are typified by overburden soils of 3 to 30 m (10 to 100 ft) in thickness. The soils are generally fine-grained, classifying as elastic silts (MH), lean clays (CL), and fat clays (CH).

The overburden soils represent the insoluble material left behind after solutioning of the rock takes place. These aluminum silicates are the by-products of the solutioning process and are typically of medium stiff to very stiff consistency. The nature of the solutioning process creates a volume reduction, which is a function of the insoluble materials in the parent rock, Belgeri and Shin (1). The leftover by-products originate as a soft paste-like material because it occupies less volume. This residual material usually lacks structure, unless formed from very sandy or shaley limestone, Sowers (4). The residual material increases in stiffness with age generally through two processes: consolidation due to the weight of subsequent deposits, and desiccation due to drying.

Within the karst residuum remain random zones of very soft material. In these zones the subsequent consolidation of the paste-like residuum is inhibited. These zones are most frequently seen where solutioning has formed a trough in the rock surface. At these locations consolidation is less likely to occur due to soil arching, wherein overburden stresses are transferred to adjacent stiffer soil zones or rock pinnacles. Also in these zones, desiccation is prevented because moisture is inevitably allowed to these areas.

Still, in many instances similar soft zones may result even though soils have been consolidated to some degree. This occurs near the rock contact following soil erosion into solution openings in the rock. This subsurface erosion, or "raveling" usually continues until the openings are plugged with residuum, stopping the circulation of water. The result is the same - a very soft paste-like material occurring randomly within an otherwise stiff soil.

The writers have found these very soft soil zones to vary from under-consolidated to slightly preconsolidated based on laboratory testing. These zones can be characterized as soft to very soft consistency with Standard Penetration Test (SPT) resistances of WOH (weight of hammer) to 3 blows per foot. The occurrence of these zones is totally unpredictable. In this sense, the residuum is as problematic for the design engineer as other karst features such as solution openings that can cause sinkhole development, or pinnacled rock. All of these features present significant design and construction challenges.

Settlement Evaluation

Using classical consolidation theory, and representative test boring and laboratory data, it is not difficult for the geotechnical engineer to accurately calculate settlements imposed by a structure. For soft soil zones, calculation of settlement is just as routine. The great difficulty is the inability to predict if and where such zones will occur.

Take for example the soil/rock profile and test boring information shown in Figure 1. Based strictly on the test boring data shown it would seem appropriate to assume stiff to very stiff fine-grained soil in the evaluation of settlement. Suppose, however, that the borings had been drilled at the nearby locations shown on Figure 2. Clearly, the evaluation and assumptions would be quite different, as probably would the resulting foundation scheme.

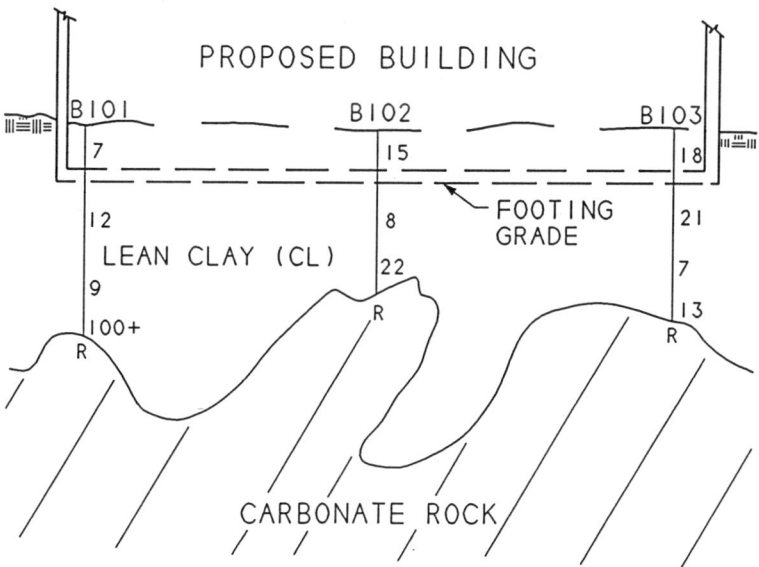

Figure 1. - Competent soils encountered within test borings.

Comparing again Figures 1 and 2, it appears that very different conclusions can be formed at the same site, depending on whether soft soil zones and solution voids happen to be encountered at test hole locations. The point is, we cannot predict where, and at what depth these soft soil zones and solution voids are present. We can plan comprehensive subsurface investigations using a variety of techniques such as borings, air track probes, cone penetrometer testing, and geophysics, but we cannot say for certain where these zones occur, except for those we happen upon.

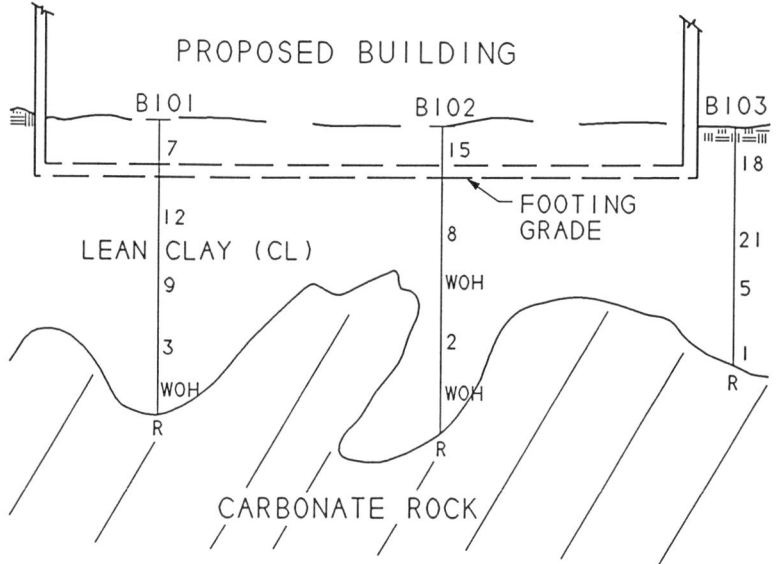

Figure 2. - Soft soils zones encountered in test borings.

This is not to say that well-planned subsurface investigations should not be undertaken, or that classical settlement analyses should not be applied. What it does mean is that there is an uncertainty that will remain regarding future performance of the structure that engineers can only hope to reduce based on their assessment of risk, but which they probably cannot completely eliminate.

Qualitative Evaluation of Risk

Regardless of whether tolerable settlements have been calculated, the foundation scheme should be selected based on at least some qualitative evaluations of the potential for future subsidence owing to undetected soft soil zones or sinkhole development. Qualitative assessment of site risk should include evaluation of various "geologic risk factors" such as the notoriety of the rock formation to produce sinkholes as may be gleaned from topographic maps, geology maps, air photos and sinkhole maps. Important also will be the assessment of specific site features such as whether ground water is well below the rock contact, and the presence of karst features such as solution voids, soft soil zones or pinnacled rock. These can be evaluated with a thorough site reconnaissance and subsurface investigation. Geologic risk factors that forebode the likely presence of undetected soft soil zones and potential for future sinkholes are shown in Table I.

Table I - Geologic Risk Factors

Likely Occurrence of Undetected Soft Soil Zones	Likely Occurrence of Future Soil Raveling and Sinkhole Development
High sinkhole density determined for rock formation Soft soil zones encountered within borings Solution voids encountered Pinnacled rock encountered	High sinkhole density determined for rock formation Ground water encountered below rock surface Soft soil zones encountered within borings Solution voids encountered Pinnacled rock encountered

The impacts of the development on the site must also be weighed. These "developmental risk factors" include: the final thickness of overburden following site grading, changes to the amount and concentration of surface water flow, ground water pumping, and blasting, DeStephen and Wargo (3). These factors, shown in Table II, are critical to assessing whether there is low, moderate, or high likelihood of future subsidence.

Table II - Developmental Risk Factors

Higher Likelihood of Settlements Due to Soft Soil Zones	Higher Likelihood of Future Sinkhole Development
Higher column loads Greater number of foundation elements Ground water pumping	Reduction of overburden thickness Ground water pumping Concentration of surface water infiltration Vibrations due to construction or blasting

When evaluating the risks associated with developing a site in karst terrain, it is also appropriate to assess the consequences of differential settlement caused by soft zones. The consequences of excessive settlement of a single spread footing in a light manufacturing building may be entirely different from the consequences of the same event in a school, nursing home, hospital or prison. The effects on the building occupants, ease and cost of repair, and the cost of temporary functional loss of a portion of the structure are some issues that can be weighed against the initial costs of constructing a lower-risk foundation.

Remedial underpinning of one or two settled spread footings in a large warehouse using compaction grout columns or pin piles may be far less costly than initially supporting all walls and columns on these special foundations. In such a case, the owner may be willing to accept a somewhat higher risk of settlement by choosing a foundation system with a lower initial cost. In contrast, the consequences and difficulties of repairing a settled column in a school, nursing home, or prison may be reason enough to choose a more conservative and costly foundation design.

Foundation Design Alternatives

Foundation alternatives must be provided that allow the owner to reduce or eliminate his risk. Various foundation alternatives applicable to karst terrain are shown in Figure 3. In this figure the foundation alternatives are shown from estimated lowest to highest cost considering both initial costs and potential for extra costs in the field. They are also shown with respect to decreasing risk of future subsidence. The figure was based on a medium-sized building such as a warehouse or school with light to moderate column loads.

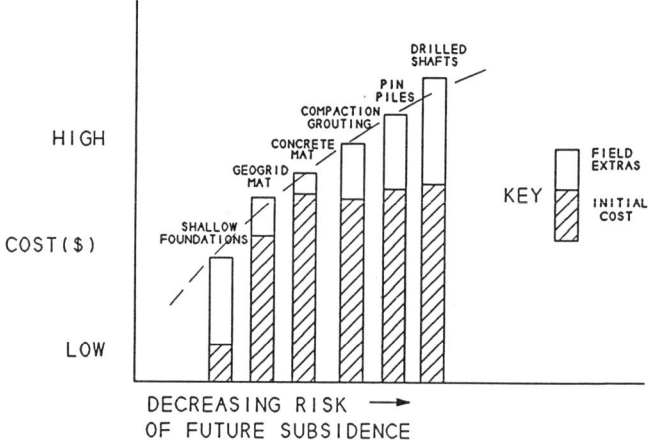

Figure 3. - Generalized cost versus Risk for foundation alternatives

Shallow foundations are used when it is calculated that the subsidence risk is low based on assessment of geologic and developmental risk factors. Shallow foundations must often be placed on a combination of soil and rock, and this in itself presents construction difficulties that can only be resolved in the field.

Excavate-replace methods are used to control estimated settlements, or as a field remediation technique where previously undetected soft soil zones have been encountered. Crushed stone is generally used, sometimes in conjunction with filter fabric, to counter the potential for future raveling.

A similar design concept is to use layers of geogrid-reinforced crushed stone to form a semi-rigid "composite mat" of higher strength material. This composite mat is generally installed to some specified depth beneath the entire building area. This is used to distribute the higher foundation stresses within the composite mat. Deeper soils may be probed for soft zones using hand augers or geoprobes during construction.

This foundation technique provides some relief from future subsidence, but does not eliminate the risk.

Rigid concrete mat foundations can be used to control settlements where small soft soil zones or small raveling zones are expected within an otherwise stiff soil mass. The lower bearing pressure and relative rigidity of a reinforced concrete mat allow it to span over small voids or soft zones and transfer the load to adjacent stiffer soils. If the likelihood of numerous or larger soft soil zones or raveling zones is moderate to high, then a mat foundation becomes more risky as it may not adequately span these areas without excessive deflections and/or settlements.

Another way to overcome karst conditions assessed to be moderate to high risk is a ground modification technique called compaction grouting. Compaction grouting uses a low slump grout under pressure to displace and consolidate soft soil zones. This allows for the use of shallow foundations where they might otherwise not be considered feasible . Compaction grouting can be considered when the ground water table is below the rock and the residual soils are not saturated. This technique both rectifies soft soil zones and serves as a prevention against future soil raveling and sinkhole formation. Preconstruction soil improvement by compaction grouting in karst areas can be a cost effective alternative to conventional deep foundations or pin pile foundations, Stapleton et. al., (5).

Pin piles are sometimes used in karst terrains, particularly where the rock surface is fairly shallow and is likely to have a highly irregular surface. Pin piles consist of small diameter, cast-in-place piles that are typically constructed using drilling and grouting equipment conventionally associated with ground anchorages. Pin piles can be designed for end-bearing or skin friction. In karst conditions, it is preferable to socket the pin piles into rock and develop capacity through the grout-rock bond. The advantages of rock sockets are that they allow the engineer to evaluate the competency of the rock based on observation of the drilling, and they transfer the load by skin friction so the engineer need not worry about bearing failure or settlement caused by undetected soil seams or voids immediately below the pile tip.

Typically the most costly foundation scheme when considering both initial costs and extra, field-related costs, is drilled shafts (caissons). However, when designed together with a structural slab, drilled shafts can virtually eliminate the risk of future subsidence problems. Drilled shafts typically cannot be successfully belled in karst areas due to perched ground water, soft soil zones prone to caving, and hard sloping rock surfaces. Besides their high initial cost, they represent a large potential for extra costs during construction. Extra costs generally occur due to inability to dewater, variable shaft lengths, and considerable rock excavation due to unsuitable bearing surfaces such as sloping rock, mudseams, and voids. Dewatering of drilled shafts can also aggravate subsurface soil erosion and possibly lead to sinkhole development, Conner and Winter (2).

Conclusions

Karst residual soils, and the carbonate rocks from which they are formed, present uncertainties with regard to future building performance which cannot be predicted. Although with classical consolidation theory, engineers can estimate settlements for a known condition, such methods are meaningless relative to the undetected random occurrence of soft soil zones, and the potential for subsurface erosion into solution voids. No longer is it justifiable to assume that settlement estimates for known conditions can be used to state predictions of overall building performance.

It is the responsibility of the design engineer to assess uncertainties relative to the future performance of the building and convey these risks to the owner. The goal is to provide alternative solutions that seek to reduce risks associated with undetected soft soil zones and solution voids.

References

1. Belgeri, J. J., and Shin, C. J., "Subsurface Conditions and Foundation Construction on Pinnacled Carbonate Bedrock," *Proceedings of the 20th Ohio River Valley Soil Seminar*, Louisville, Kentucky, 1989, pp.2-3.

2. Conner, S.E. and Winter, S. J., "Sinkhole Remediation Beneath Existing Building, Pulaski County Courthouse Addition," *Report to Pulaski County*, Pulaski, Virginia, 1994, pp. 1-3.

3. DeStephen, R. A., and Wargo, R. H., "Foundation Design in Karst Terrain" *Bulletin of the Association of Engineering Geologists*, Vol. XXIX, No. 2, 1992, pp. 165-173.

4. Sowers, F.G., "Settlement in Terrains of Well-indurated Limestones," *Proceedings, Analysis and Design of Building Foundations*, Lehigh Univ., Bethlehem, Pa., 1975, pp. 701-718

5. Stapleton, D.C., Corso, D., and Blakita, P.M., "A Case History of Compaction Grouting to Improve Soft Soils Over Karstic Limestone," *Proceedings of the Fifth Multidisciplinary Conference on Sinkholes*, 1995.

Applications of Soil Nailing in Residual Soil

James W. Sigourney[1], Member, ASCE

Abstract

Soil Nailing is a form of in situ ground reinforcement used to strengthen a soil mass. Typically, steel rods or "nails" are drilled and grouted in a cut face using a closely spaced pattern. The nails reinforce and strengthen the soil mass. Due to excavation, the ground movement is limited by the passive resistance of the soil nails. A shotcrete facing is commonly used to maintain the soil near the face between the nails. Drainage is very important when using this method of ground support. When used for excavation support or slope stabilization, soil nailing can have advantages that are unique to residual soils. There are some disadvantages as well. This paper looks at both and hopes to advance the use of soil nailing in residual soils.

Introduction

Residual soils are the result of the physical and chemical weathering of rock in place. Most weathering profiles result in a relatively impermeable residual soil that exhibits sufficient cohesion to stand vertically for a five-foot excavated cut. These characteristics are particularly suited to soil nailing. An overview of soil nailing will be provided first. Applications of soil nailing for the three representative cases in residual soils will then be discussed. Limitations of the use of soil nailing in residual soils will then be defined.

---

[1] Chmn. of the Board and Pres., TerraTech, Inc.
101 Loudoun St., SW., Leesburg, VA 22075

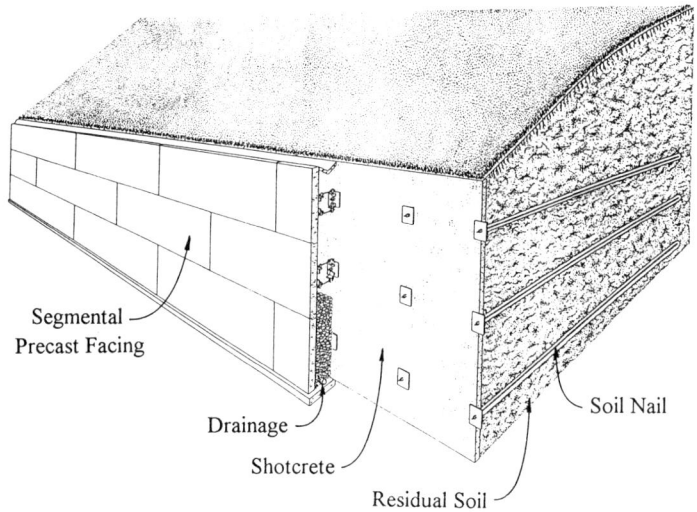

Figure 1. Soil Nail Retaining Wall

Soil Nailing

Soil Nailing is a form of ground modification that reinforces a soil mass during excavation from the top down using an array of nails. Three major applications include retaining walls, slope stability and temporary excavation retention systems.

The nails are usually constructed using steel rods grouted into drilled holes that are typically inclined ten to twenty degrees from horizontal. Soil nail spacing seldom exceeds five feet vertically or horizontally. The nails must be long enough to create a stable mass. They are typically sixty to 85 percent of the height of the excavation if the surface at the top of the wall is level and no major surcharge is present. The typical construction procedure begins with the first lift of excavation. The face of the excavation is supported by shotcrete reinforced with wire mesh. Soil nails are then drilled and grouted. The order of the last two steps is often reversed. These steps are repeated for each lift until the excavation is completed.

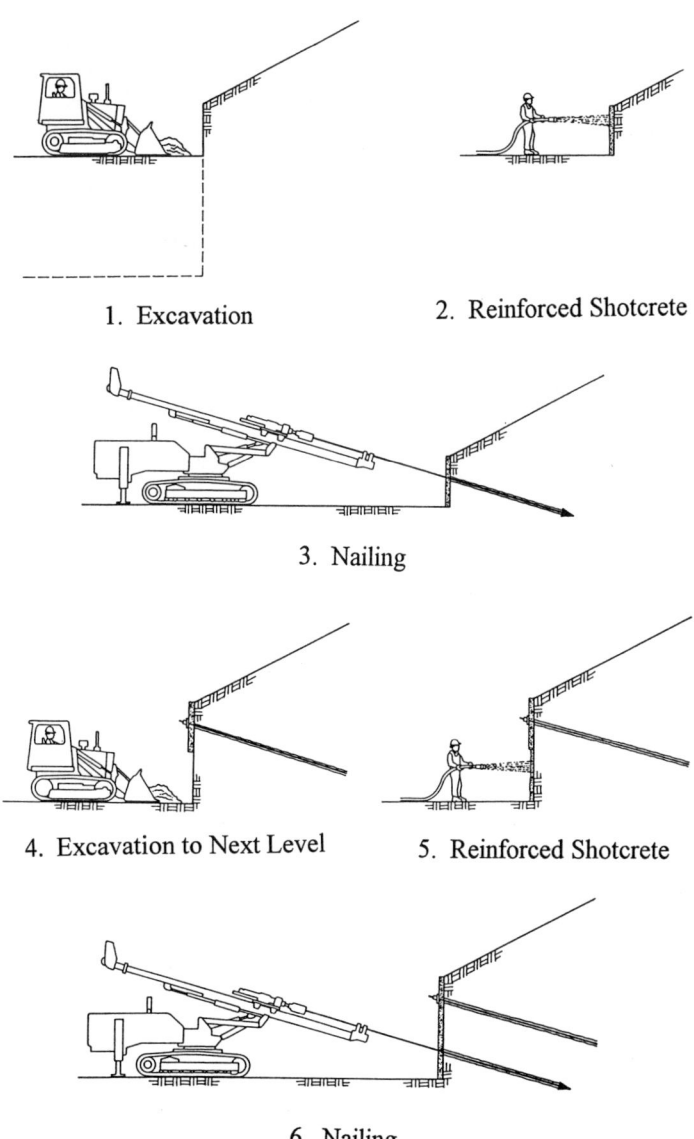

Figure 2. Construction Method

Several design procedures are currently being used for the analysis of soil nailed structures. Potential failure surfaces through the soil mass are checked. Some only consider the tensile capacity of the nail when assessing the stability of the soil nailed mass, including the Davis method described by Shen et. al. (1981), and methods proposed by Gassier and Gudehus (1982). Schlosser (1983) developed stability analysis that allows for the bending stiffness and the shear capacity of the nails. A mathematically rigorous method using a log-spiral surface has been proposed as the "Kinematical" Limit Analysis Approach for the design of soil nailed retaining structures (Juran, et. al. 1990). The Kinematical limit design approach provides an estimate of maximum tension and shear forces mobilized. The industry is continuing to debate and develop methods, (i.e. Caltrans, Golder, Jewell, Plumelle). What should be noticed is that similar results are obtained from all these methods for normal design conditions that are vertical walls without slope surcharge. Circular, three-part wedge, and bi-linear failure surfaces predict similar factors of safety except for larger values of soil nail adhesion values The one exception is that some methods underestimate the required tensile capacity of the nail. As a result, an empirically derived "apparent earth pressure envelope" is used to check the nail tensile forces.

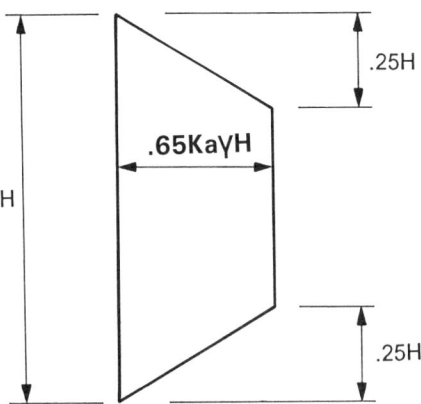

Figure 3. Apparent Earth Pressure Envelope

The maximum stress in the nail is not at the face of the wall, but defines a boundary between an active and a resistant zone. This is similar to the boundary of the active wedge for a mechanically stabilized embankment achieved by reinforcing the backfill with metal strips. The modified Davis method appears to provide the most accurate results defining this boundary when compared to test results. In practice, computers are required for the detailed computations.

Each engineer must use his experience in choosing the adhesion value of the grouted nail body to the soil. A representative number of soil nails are tested in the field to verify the tensile capacity of the nail.

Ashville, North Carolina

This project required widening and realignment of approaches to Bridge No. 76 over Swannanoa River. The construction of a retaining structure is required to widen the approach. A soil nailing retaining wall was selected.

Residual soils are present in the cut sections for the wall. The residual soil is dominantly fine, sandy, silty saprolite. Extensive interlayers of non-cohesive, silty sands are common. It is typical to encounter a residual silty clay at the ground surface. The residual soils are slightly micaceous overall. Groundwater is found below the excavation level for the soil nailed wall and no water problems are anticipated. The design height of the wall will be up to thirty-three feet.

For stability analysis of the wall, a friction angle of thirty-three degrees, a unit weight of one hundred twenty pounds per cubic foot, and a cohesion value of zero was used. The PH of the soil ranges from 7.4 to 7.6. The restivity of the soil ranges from 24,800 ohm-cm to 38,900 ohm-cm. The soil inclination at the top of the wall is twenty-seven degrees from horizontal. The soil nailed wall design uses a segmental precast facing with a fractured fin architectural finish.

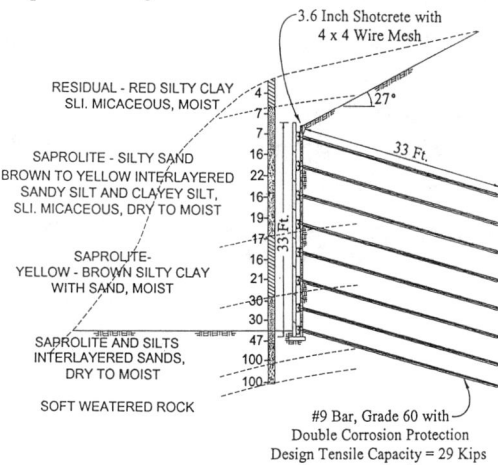

Figure 4. Section Through Soil Nailed Wall

Reading, Pennsylvania

Clearance improvement under existing bridges for Conrail trains. To lower the railroad track level, excavation below the base of existing retaining walls was required. Soil nailing was selected to act as a retaining structure to underpin the existing walls. The earth below the walls was highly variable and included residual soil, weathered rock, and rock. The residual soil was derived from the weathering of Carbonate Rocks, principally Limestone.

The soil is generally clayey. The boundaries between the rock and soil are unpredictable.

Soil nailing was selected to adapt to the random mix of rock and soil. The limited easement also favored soil nailing. The soil nailed design uses vertical drains between the nails and a reinforced shotcrete finish.

Figure 5. Applying Shotcrete

Figure 6. Installing Soil Nails

Pittsburgh, Pennsylvania

The widening of highways creates the demand for excavating steeper slopes. Frequent slides have occurred and a combination of rock bolting and soil nailing is used to stabilize the slopes. In the Pittsburgh area, the bedrock consists of cyclic sedimentary rock of the Conemaugh Formation. This formation typically has alternating layers of Birmingham Shales, Pittsburgh Red Beds, Sandstone, Limestone, and Coal.

The Red Beds are principally consisting of claystones, characterized by bright red coloring and rapid decomposition at the outcropping. This results in residual soils derived entirely from the weathering of the claystone. Landslides are frequently observed at the contact zone between the residual soils and bedrock. Both shotcrete and welded wire mesh have been used as a facing between nails.

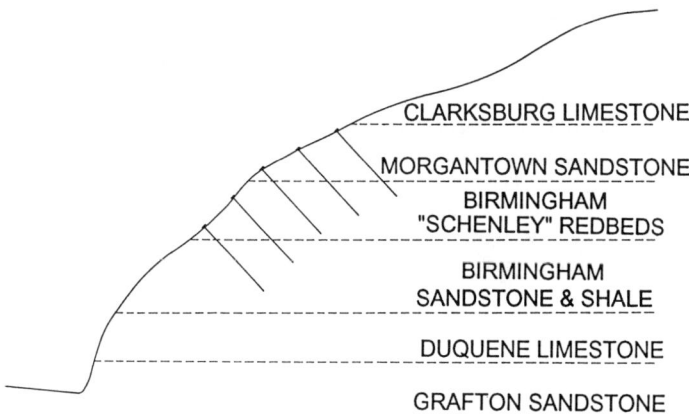

Figure 7. Rock Formation

Limitations

Many landslides occur in residual soils. Most occur during the excavation of a slope. Soils derived from shales and claystones seem the most susceptible to landslides. The most devastating landslides occur after a rainstorm which adds to the external force on the soil mass, reduces the shear strength of the soil and changes the pore pressure.

Soil nailing is a form of ground modification that can prevent landslides; however, it cannot accommodate large quantities of water. Groundwater drainage systems must be constructed. A shotcrete facing is the most susceptible to icing and has a limited effectiveness of drainage. The most satisfactory method for drainage uses a drainage stone placed behind segmental precast facing. Slope failure outside of the reinforced soil mass must also be checked.

Some soils should be avoided. Soil nail capacity cannot be effectively developed in cohesive soils with a liquidity index greater than 0.2 and may be susceptible to creep. Fine dry cohesionless sands cannot stand vertically long enough to use soil nailing. Any porous soil with groundwater flowing should be avoided. Fill materials must always be suspect for using soil nailing. Any soil with an N value less than ten blows per foot should be avoided.

Some physical limitations must also be considered. Underground easements may be required. Utility conflicts may also preclude the use of soil nails.

Conclusion

The normal weathering of rock to create residual soil typically provides and excellent material to be reinforced with soil nails. The internal resistance against sliding during excavation is achieved by reinforcement with nails. The mixture of residual soil and rock often makes the installations of other retaining structures too costly. It is, however, ideal for the installation with soil nailing equipment. The purpose of this paper is to increase the understanding and application of soil nailing in residual soils.

References

Gassler, G. and Gudehus, G. (1981), "Soil-Nailing - Some Soil Mechanical Aspects of In-Situ Reinforced Earth", Proceedings of 10th International Conference on Soil Mechanics and Foundation Engineering, Vol. 3, Session 12, Stockholm

Juran, I., Baudrand, G., Farrag, K., Elias, V., (1990) "Kinematical Limit Analysis for Design of Soil-Nailed Structures", Journal of Geotechnical Engineering. Vol. 116, No. 1.

Schlosser, F. (1983), "Analogies et differnces dans le Comportement et le Calcul des Ouvrages de Soutenment en Tene Armee et par Clouge du Sol", Amales de L'institut Technique du Batiment et des Travaux Publics No. 418

Shen, C.K., Herman, L.R., Romstad, K.M., Bang, S., Kim, Y.S., and Denatole, J.S. (1981), "An In-Situ Earth Reinforcement Lateral Support System", Report #81-03, University of California, Davis, California, March.

Estimation of In Situ Hydraulic Properties of Saprolite

Gordon M. Matheson, M. ASCE[1]

Abstract

Saprolite is a soil-like material developed from the chemical alteration of rock that has the relic texture and structure of the rock still visible. Saprolite developed above metamorphic and igneous rocks are common in the Piedmont of the eastern United States. Due to saprolite's gradational nature between soil and rock, design hydraulic properties are difficult to assess. In fully developed soil, primary porosity controls the flow of water and contaminants. In rock, water and contaminant flow is controlled by secondary porosity along rock fractures since primary porosity is negligible. Saprolite typically contains some primary porosity as well as relic secondary porosity and thus represents a combined flow system.

Hydraulic properties of saprolites are also poorly understood. The primary effective porosity of saprolite is typically in the range of 25 to 35 percent. However, the conductivity of the soil matrix is relatively low, thus water production and flow velocity through the soil matrix is also low. If relic fractures are present, the secondary porosity controls water flow. This porosity may be low (less than 5 percent), yet the conductivity along the fractures may be relatively high. Thus, a flow volume equivalent to a porous media can occur in saprolite, however, the flow velocities will be higher.

The hydraulic conductivity of saprolites is most commonly measured by small volume displacement tests such as "slug tests". Recent work by Rovey and Cherkauer (1995) has shown that the hydraulic properties of soil and rock are typically underestimated by slug tests due to the size effect from small to large volumes. This can result in the underestimation of hydraulic properties of saprolites by as much as one half an order of magnitude.

---

[1] Principal, Schnabel Engineering Associates, 10215 Fernwood Road, Bethesda, Maryland 20817

Incorrect assumptions of these properties can have significant design implications. For example, underestimation of hydraulic conductivity and overestimation of the porosity that controls flow at environmental sites can cause a significant underestimation of contaminant travel time. At sites where dewatering is required, the measurement of conductivity controlled by relic fractures combined with porosity estimated from primary porosity may lead to calculation of greater dewatering volumes and lower drawdowns and drawdown rates than are experienced in the field.

Introduction

The Piedmont of the east coast of the United States is underlain by a broad band of metamorphic and igneous rocks. This band consists primarily of schist, gneiss, phyllite and granitic rocks. These rocks are composed of several unstable minerals that alter and cause the rocks to chemically decompose. This decomposition gradually changes the rock into soil. A transition zone is present between sound rock and soil that is termed saprolite. Saprolite typically maintains the structure and texture of the original rock.

Due to the gradational nature of these materials their hydraulic and geotechnical properties are quite variable. Typically, the parent rock has little primary (intergranular) porosity and the flow of ground water is controlled by flow through fractures or secondary porosity. Completely disintegrated rock or soil hydraulic properties are controlled by primary porosity flow since the relic secondary porosity has typically been destroyed. Saprolite has the combined properties of the soil and the rock. Misunderstanding of the flow mechanics and characteristics of saprolite can lead to erroneous design assumptions. The purpose of this paper is to discuss the concept of flow in these materials and present data on typical hydraulic properties measured in these materials.

Saprolite Hydraulic Properties

The disintegration of rock typically creates a saprolite that gradually becomes firmer with depth. For purposes of this discussion, sound rock is assumed to be represented by material that has a Standard Penetration Test blow count of 100 blows for two inch of split spoon sampler penetration or less. This is a somewhat arbitrary selection and has been found through experience to represent where rock can be cored and relatively good core recovery can be obtained.

The upper boundary between saprolite and soil is less well defined. Based on experience in the mid-Atlantic states with schists, phyllites and granitic rocks, saprolitic material that maintains the relic texture and structure of the underlying rock begin at

Standard Penetration Test blow counts between 20 to 30 blows per foot. Thus, the saprolite zone under consideration is considered to be bounded by the Standard Penetration Test blow count ranges specified (i.e., 20 to 30 blows per foot to 100 blows per two inch split spoon penetration).

Within this zone the saprolite hydraulic properties will vary. In the stiffer zones the relic fracture structure (secondary porosity) will control the hydraulic properties while in the upper zones, the primary porosity will control the hydraulic properties.

The primary porosity of the saprolite is typically quite high. Based on density testing on undisturbed samples, the porosity of saprolite ranges from about 30 to 40 percent. Since effective porosity controls actual water flow, it is reasonable to assume that this porosity is between 25 to 35 percent. Secondary porosity of the rock is typically low. In highly fractured zones this porosity may approach 5 percent, however, more typical values are in the range of 0.5 to 2.0 percent when taken on a rock mass basis. Stiffer saprolites will have a porosity close to rock.

When deciding what porosity should be used in hydraulic calculations, the designer must decide whether the hydraulic conductivity of the zone of concern is controlled by the primary or secondary porosity. Thus, the estimation of hydraulic conductivity of the mass and the assessment of the flow mechanics become important in this process.

Measurement of Hydraulic Properties

Four zones of hydraulic conductivity are present in a saprolite profile. These are:

- o The upper most zone, dominated by primary porosity, has the characteristics of a porous media.

- o An intermediate zone that consists of saprolite with relic rock fractures, is often controlled by the conductivity of the relic fracture systems.

- o The rock system is controlled essentially entirely by the secondary fracture porosity.

- o An intermediate zone at the base of the saprolite zone, immediately above and below the saprolite and rock interface. In this zone a layer of lower hydraulic conductivity than the overlying more soil-like material and the underlying rock is often present. Within this zone the hydraulic conductivity due to primary porosity is relatively low due to the incomplete disintegration of the altering minerals. The secondary porosity is present but this is also of low conductivity since many of the fractures are clogged by either clay minerals from disintegrating feldspars or chemical precipitates (typically iron or manganese oxides or carbonates) that have been deposited by ground water.

In saprolite terrains a perched water table is often seen immediately above the rock surface and extending several feet into the rock, that is in poor hydraulic connection to the underlying bedrock aquifer. The low conductivity zone delays infiltrating water from reaching the bedrock aquifer.

Piezometers placed in the saprolite compared to piezometers placed in the bedrock often show a significantly different potentiometric surface. This creates a vertical hydraulic gradient that partially causes the "delayed yield" effect seen in long-term aquifer tests of the bedrock aquifer. Depending on the location of the monitoring point the vertical gradient may be either upward or downward. It should be noted that in some rocks that the secondary porosity fractures are either not present, are not persistent or are not well connected in the rock mass. In these areas the rock mass may be lower in hydraulic conductivity than the overlying saprolite. It is the authors experience that this condition is relatively rare in the Piedmont rocks.

The measurement of hydraulic conductivity of saprolites is a difficult task. For most geotechnical investigations either laboratory grain size distribution correlations or relatively small scale borehole "slug" or "falling head" tests are completed to measure hydraulic conductivity. Seldom are aquifer pumping tests completed.

It has been found that the laboratory and small scale borehole tests generally underestimate the larger scale hydraulic conductivity that controls flow during construction. The general nature of this problem was described by Rovey and Cherkauer (1995). They found that the hydraulic conductivity was related to the volume of the mass tested and that, especially in materials whose properties are controlled by secondary porosity, the larger the mass tested the higher the hydraulic conductivity. They found that for materials with only primary porosity that slug tests will often under estimate hydraulic conductivity by a factor of three. For materials controlled by secondary porosity slug tests were found to underestimate large scale conductivity by factors ranging from 2 to 500. It should be noted, that the upper bound of this range was found in highly porous carbonate terrains and does not apply for saprolites discussed in this paper. It general, it appears that the small scale tests will underestimate hydraulic conductivity by a factor of between 2 and 5. This range was recently confirmed for a site in the Washington D.C. area where slug tests indicated a hydraulic conductivity of a factor of about 2.5 less than a subsequently completed long-term aquifer test. Laboratory estimates based on grain size distributions were found to be up to a factor of 10 lower than large scale tests.

Typical values for the in situ hydraulic conductivity of saprolite are presented in Table 1 and represent values from testing completed in the mid-Atlantic states. Table 1 also presents the parent bedrock type, Standard Penetration Test blow count range for the test zone and type of test completed. It can be seen from this table that saprolite hydraulic conductivity will typically fall within a range of about 0.1 feet per day (.000035 cm/sec) to 5 feet per day (.0000176 cm/sec).

Table 1
Representative Results of
Borehole Hydraulic Conductivity
Testing in Saprolite

Rock Type	Standard Penetration Test Blow Count	Type of Test	Measured Hydraulic Conductivity (ft/day)
Phyllite	28 to 100/6"	Falling Head	1.3
Phyllite	52 to 100/4"	48 Hour Pump Test	5.8
Phyllite	30 to 100/1"	Slug Test	0.4 to 4.0
Gneiss	100/6"	Slug Test	0.1
Gneiss	100/3"	Slug Test	0.9
Schist	36	Slug Test	0.1
Schist	77	Slug Test	0.1
Schist	57 to 100/3"	Slug Test	0.2
Schist	136 to 100/6"	Slug Test	1.1

Implications for Design

The proper selection of properties for saprolites can have significant design implications. Potential important areas of impact include estimates of contaminated ground water travel time and estimates of ground water dewatering for construction or remediation.

Contaminant travel time is controlled largely by the ground water flow velocity. The ground water flow velocity is calculated by use of the familiar expression:

$$\overline{v} = Ki/n_e$$

Where:

$\overline{v}$ = ground water average linear flow velocity
K = hydraulic conductivity of material
i = hydraulic gradient of ground water table or potentiometric surface
n_e = effective porosity of material

The selection of the hydraulic conductivity and porosity are critical in this calculation. The hydraulic gradient is important and is normally calculated from a ground water contour map based on wells installed on a site.

A common problem in saprolites is the use of the hydraulic conductivity measured from slug tests and the primary porosity of the saprolite in this calculation. For example, if the ground water gradient is assumed to be .005 (a value typical of the Piedmont), the hydraulic conductivity is assumed to be 1.0 feet per day (.00035 cm/sec), and the primary effective porosity is assumed to be 25 percent, the implied ground water flow velocity is .02 feet per day (.006 m/day) or 7.3 feet per year (2.2 m/year), an intuitively slow flow rate. If the secondary porosity is actually controlling the ground water flow this calculation underestimates the ground water flow velocity. Using an average secondary porosity of two percent, the calculated ground water average linear velocity is 0.25 feet per day (.077 m/day) or 91.25 feet per year (28.1 m/year), a 12.5 times increase in velocity. It could be argued that the relic fractures in the saprolite are not oriented in the direction of flow so the actual average linear velocity is slower, and there is validity to this argument. It is clear, however, that significant non-conservative error can be introduced into this calculation by not understanding the mechanics of ground water flow through the saprolite. This is further complicated by the potential non-conservative error introduced by the use of small scale borehole tests to estimate in situ hydraulic conductivity. If the potential error is a factor of 2 to 5 as found by Rovey and Cherkauer (1995), then the average linear ground water flow velocity could be underestimated by a factor of 25 or more.

Dewatering is another common concern in saprolites. If these materials are considered as primary porosity controlled media, then dewatering of the saprolites would be difficult. Figure 1 presents a calculation using a Theis drawdown assumption for a 50-foot (15.3 m) deep pumping well in saprolite using the same range of porosity and hydraulic conductivity stated in the previous paragraph. If the well is assumed to be controlled by primary porosity then the drawdown at, say, 30 feet (9.2 m) from the well is found to be about 8 feet (2.5m) after 30 days. If secondary porosity controls the dewatering rate then the drawdown is about 15 feet (4.6m) at the same distance and time or about 87.5 percent more drawdown. This shows that the dewatering of saprolites is not as difficult as may be envisioned based on a primary porosity controlled assumption. This has been the authors experience in the mid-Atlantic region.

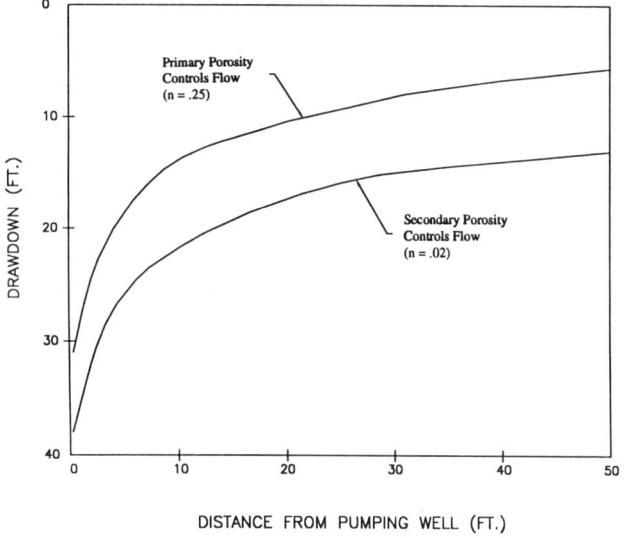

Figure 1: Comparison of Predicted Saprolite Drawdown for Primary and Secondary Porosity (10 gpm discharge, K = 1.0 ft/day, 30 days of pumping)

Conclusions

This paper has presented a review of the hydraulic properties and flow mechanics of saprolites. It is important to recognize that saprolites have relatively unique properties due to their gradational nature between soil and rock. Based on the information provided in this paper the following conclusions can be drawn:

o   Saprolites are gradational between soil and rock and possess many properties of both materials. It is important to recognize that secondary or relic fracture porosity is likely to control ground water flow through much of these materials. The secondary porosity is typically in the range of 0.5 to 2 percent as compared to primary effective porosity of 25 to 35 percent.

o   Saprolite can generally be recognized in the field by their standard penetration test blow count values. Experience and observation have shown that soils will retain significant relic texture and fracture structure at blow counts of 20 to 30 blows per foot and greater. Sound rock is defined in this paper as a Standard Penetration Test blow count of 100 blows for two inch of penetration or less.

- Small scale borehole tests and laboratory gradation correlations to estimate hydraulic conductivity of soil and rock have been shown by others to significantly underestimate the larger scale mass hydraulic conductivity that is important in most engineering design. For materials controlled by secondary porosity, this underestimation is typically in the range of 2 to 5 times the actual value.

- Use of primary porosity in calculation of ground water average linear velocity in saprolite can result in a significant underestimation of the rate of ground water flow. Including the potential impacts of underestimation of hydraulic conductivity, the ground water flow average linear velocity could be underestimated by 25 times the actual flow velocity or more.

- The use of primary porosity in calculation of dewatering requirements for excavations in saprolite can result in an underestimation of drawdowns from the dewatering. This can result in an excess number of wells being designed for a dewatering system with associated higher construction costs.

Saprolites, by their unique nature, require careful assessment during the design process. Further research into the relative hydraulic conductivity of the saprolite matrix in various stages of decomposition as compared to relic fracture conductivity and frequency would be of benefit to the engineering community.

Reference

1. Rovey, Charles W., and Cherkauer, Douglas S., 1995. Scale dependency of hydraulic conductivity measurements. GROUND WATER, Vol. 33, No. 5, September-October

SUBJECT INDEX

Page number refers to the first page of paper

Compaction, 37
Compaction grouting, 49
Construction, 57
Construction materials, 37

Dam construction, 27
Dams, embankment, 27
Design, 57
Drainage, 57
Drains, 27

Excavation, 57

Fatigue tests, 12
Foundation design, 49

Geology, 1, 27

Hydraulic conductivity, 66
Hydraulic properties, 66

In situ tests, 37

Jar tests, 12

Land usage, 1
Limestone, 49

Measurement, 37
Mineralogy, 1, 57, 66

Oil shale, 37

Permeability, 27
Permeability, soils, 66
Predictions, 12
Pressuremeters, 37

Residual soils, 1, 49, 57, 66
Risk, 49
Rock fills, 37
Rock properties, 37
Rocks, 1, 66

Seepage, 27
Shake table tests, 12
Shales, 12
Shear strength, 27
Shotcrete, 57
Sinkholes, 49
Soil mechanics, 27
Soil nailing, 57
Soil porosity, 66
Soil settlement, 49

Thickness, 1

Weathering, 1, 12

AUTHOR INDEX
Page number refers to the first page of paper

Clift, Anne Eckert, 37
Conner, Steven E., 49

DeStephen, Raymond A., 49

Koncagül, Engin C., 12

Martin, Ray, 27
Matheson, Gordon M., 66

Pavich, Milan J., 1

Santi, Paul M., 12
Sigourney, James W., 57

Wilson, Chuck, 27

AVAILABLE BOOKS OF THE 1996 ASCE ANNUAL CONVENTION & EXPOSITION

WASHINGTON, DC ★ NOVEMBER 10 - 14, 1996

Proceedings of sessions held in conjunction with the
1996 ASCE Convention in Washington, DC

Analysis & Design of Retaining Structures against Earthquakes
Shamsher Prakash, Editor
Geotechnical Special Publication No. 60
ISBN 0-7844-0206-X

Case Histories of Geophysics Applied to Civil Engineering and Public Policy
Paul Michaels and Richard Woods, Editors
Geotechnical Special Publication No. 62
ISBN 0-7844-0208-6

Civil Engineering History: Engineers Make History
Jerry R. Rogers, Donald Kennon, Robert T. Jaske, and Francis E. Griggs, Jr., Editors
First National Symposium on Civil Engineering History sponsored by the ASCE & the US Capitol Historical Society
ISBN 0-7844-0209-4

Civil Engineers Influencing Public Policy
Maureen K. Cotton, Editor
Sponsored by the Construction Division
ISBN 0-7844-0204-3

Design with Residual Materials: Geotechnical and Construction Considerations
Gordon Matheson, Editor
Geotechnical Special Publication No. 63
ISBN 0-7844-0207

Engineered Contaminated Soils and Interaction of Soil Geomembranes
Jay N. Meegoda, Luis E. Vallejo, and L. N. Reddi, Editors
Geotechnical Special Publication No. 59
ISBN 0-7844-0213-2

Materials for the New Millennium
Ken P. Chong, Editor
Proceedings of the Fourth Materials Engineering Conference sponsored by the Materials Engineering Division
ISBN 0-7844-0210-8

Measuring and Modeling Time Dependent Soil Behavior
Thomas C. Sheahan and Victor N. Kaliakin, Editors
Geotechnical Special Publication No. 61
ISBN 0-7844-0205-1

Non-Aqueous Phase Liquids (NAPLs) in Subsurface Environment: Assessment and Remediation
Lakshmi N. Reddi, Editor
Proceedings of the Specialty Conference sponsored by the Environmental Engineering Division
ISBN 0-7844-0203-5

For pricing and availability, contact: **AMERICAN SOCIETY OF CIVIL ENGINEERS INTERNATIONAL HEADQUARTERS** 1801 Alexander Bell Drive, Reston, VA 20191-4400 ★ Phone: 800.548.2723; 703.295.6029 (international) ★ Internet email: marketing@asce.org